C0-ATH-118

Comeuppance

Costly Signaling, Altruistic Punishment, and Other Biological Components of Fiction

William Flesch

HARVARD UNIVERSITY PRESS
Cambridge, Massachusetts
London, England
2007

Printed in the United States of America

Library of Congress Cataloging-in-Publication Data

Flesch, William, 1956–
Comeuppance : costly signaling, altruistic punishment, and other biological components of fiction / William Flesch.
p. cm.
Includes bibliographical references and index.
ISBN-13: 978-0-674-02631-5 (alk. paper)
ISBN-10: 0-674-02631-4 (alk. paper)
1. Fiction—Psychological aspects. I. Title.

PN3352.P7F53 2007
808.301′9—dc22 2007021761

To Laura, Daniel,
and Julian
(age six, who told his parents proudly,
after rebuking someone for parking at a fire-hydrant,
"*I was doing altruistic punishment!*")

Contents

Preface

In this book I cite some recent work in evolutionary psychology to try to analyze some of the biological conditions of possibility of narrative and in particular of fiction. I, too, am suspicious of evolutionary psychology. I would be glad to think, therefore, that the sometimes mildly technical theoretical and psychological arguments that I make about the nature of fiction will be judged by the insights they yield with respect to particular narratives. Readers should feel free to begin with Chapter 4, which focuses on two exemplary narratives. Chapter 4 should be pretty clear even without the earlier terminological explanation that the first three chapters provide; if you find it at all convincing you can turn to the first three chapters for an argument as to why evolution might have formed us to respond to *Oliver Twist* and *King Lear* as we do. Chapter 4 is on vindictiveness and vindication in narrative, and it shows how the basic scheme I argue for functions in these two examples. That scheme, much simplified, is this: narratives tend to contain or at least to suggest the possibility of three basic figures (though there may be more or fewer than three characters who instantiate them): an innocent, someone who exploits that innocent, and someone else who seeks to punish the exploiter. Humans are endowed by our evolutionary heritage with a propensity to punish those who cheat the innocent and with a propensity to cheer on other punishers. This is why we dislike villains and root for heroes. The biological origin of this propensity is part of what has come to be called the "evolution of cooperation," which provides the insights that are central to this book.

Every book testifies to the evolution of cooperation. It does so officially in some of its paratextual apparatus: Dedication, Acknowledgments, Notes, and Works Cited. The other elements of the apparatus appear in due order; here let me acknowledge a few of the people who made this book possible through their help and occasional severity (since this book is in part a celebration of severity):

Marshall Brown, John Burt, Patricia Chu, J. T. Daly, Veronica Davidov, Alma Flesch, Michael Gilmore, Hannah Ginsborg, Eugene Goodheart, Deborah Gordon, Nick Halpern, Neil Hertz, Ann Hochschild, Karin Lewicki, Robin Feuer Miller, Richard Moran, Paul Morrison, Tonnya Norwood, Jeff Nunokawa, Leah Price, Greg Petsko, Christopher Pye, Laura Quinney, Tom Reinert, Jamie Robbins, Adam Rutledge, James Schwartz, Steven Shaviro, Paul Solman, John Sutherland, Daniel Warren, Lindsay Waters, Caroline Wingolf, Lisa Zunshine, and the two readers for Harvard University Press.

Comeuppance

Com'esser puoto ch'un ben, distributo
più posseditor, vaccia più ricchi
di sé, che se da pochi è posseuduto? . . .

E quanta gente più là su' s'intende,
piùv'è da bene amare, e più vi s'ama.
e come specchio l'uno a l'altro rende.

—*Purgatorio*, XV, 61–63, 72–75

How can the distribution of a good
among a greater number make each richer
than for the few their own possession would? . . .

The more the souls intent on souls above,
the more they love, their loving all the better,
like mirrors shining back each other's love.

—*Purgatorio*, XV, 61–63, 72–75

The vanity in the desire to be seen as having been freely granted the evident signs of the predilection of the one we love is in fact derived from love itself, from the need to represent ourselves to ourselves and to others as loved by the person we love so much.

—*Le Côte de Guermantes*, 2. 460–461

Introduction

Our practices do not merely exploit our natures, they express them.

—P. F. Strawson, "Freedom and Resentment," p. 93

A creature cannot have thoughts unless it is the interpreter of the speech of another.

—Donald Davidson, *Essays on Truth and Interpretation,* p. 157

This book is an attempt to use evolutionary psychology to account for the surprising fact that humans can become so emotionally absorbed in stories we know to be fictions. It is addressed especially to readers who are wary of such explanations: I am wary of them myself. Some recent ideas in evolutionary biology are as beautiful, striking, subtle, and surprising as any of the big ideas in philosophy and literary theory over the last generation. But the people tempted to apply evolutionary psychology to the explanation of literature tend to be extremely reductive. (See, for example, Brian Boyd's recent polemic in *The American Scholar;* Lisa Zunshine is an eloquent exception to this rule.) Although they would certainly not put it this way, they think they have good reasons to suppose that literature cannot be as subtle and as deep as the best literary criticism takes it to be, or rather they think there is no good reason to suppose that literature could be as subtle and deep as literary criticism claims. If literature is the product of human minds that have evolved to meet biological constraints, the literature described by the best literary theory and philosophy seems impossible. The best literary critics seem to agree with the general assumption of evolutionary psychologists that their approaches are mutually exclusive. I don't think they are, and in this book I try to show how these two approaches can be combined.

If recent developments and debates in evolutionary biology are fascinatingly beautiful, it is also true, alas, that the general run of evolutionary explanations of literature and literary experience are programmatically simplistic. Evolutionary psychologists tend to fall prey to two kinds of reductionism when they talk about literary issues: either that literature's function is fairly simple or that it is biologically fairly trivial.

The first claim, naturally, leads to more analysis than the second and is therefore more widely represented in evolutionary literary criticism. This approach accounts for our capacity for literary experience (and for our experience of art in general, perhaps), by identifying something it calls narrative or storytelling or literary activity more generally. It then tries to give an explanation for how such behavior might have been selected for during what's called the Environment of Evolutionary Adaptation, or EEA. The question such an approach seeks to answer is this: why did storytelling (and its necessary complement, story-attending, being an audience for a story) evolve? The problem with this way of framing the question is that it tries to explain a complex behavior or activity (or complexly interwoven set of behaviors and activities) as a single entity, an entity that would have evolved to be the way it is because it is a useful adaptation (where useful means: because it provides an advantage in the competition for fitness or reproductive success). I don't think this account of literature as a single, evolved phenomenon or behavior is true. A capacity for narrative may *be* a useful adaptation, but that's not why it evolved.

Evolutionary psychologists in this line, which we may see as centered on the work of the sociobiologist E. O. Wilson, are tempted to look for archetypical patterns of narrative and character (of a more or less Jungian sort) to suggest that it is these that most comport with what evolved in the EEA and most resonate in our own enjoyment of the arts (E. O. Wilson 1998: 244). Criticism of this sort, from Vladimir Propp to Christopher Booker's mammoth work *Seven Basic Plots*, may give an adequate account of the simplicity but not of the complexity of literature, of why we might have certain aesthetic responses like those that fairy tales or ballads give us, but not perhaps why we have other responses that we might call more "sophisticated." (Frye [1957] is the great rule-proving exception.) Although I shall make some basic claims about narrative and narrative situations, they won't be about archetypal stories or patterns of storytelling. Wilson is suspicious of literary critical *thought*, though he likes critical observation and is a good observer himself. The criticism he has successfully promoted among his literary-minded followers too often settles only for his eloquent vapidity. The higher cognitive dimensions of art, what it is that might *fundamentally* distinguish Kafka from the Grimms, or show that there is no royal road to Kafka even if there is one to the Grimms, he abjures. In particular, in the example he uses from *Paradise Lost* (the comparison of Eden to "that fair field / Of Enna, where Proserpine gathering flowers, / Herself a fairer flower, by gloomy Dis / Was gathered") he ignores the question why Milton might be able to count on our emotional response to a myth conspicuously described as *fictional.*

As I hope will become clear, however, little that I say is necessarily inconsistent with the positive views of Wilson's school; it's just that there is much more to be explained than those views do, and much more by way of explanation that can be offered.

There is one important inconsistency between the perspective that I take and Wilson's. Some evolutionary psychologists insist on the adaptationist claim for the meaning of art and literature because they have got religion, what Wilson's bitter and bitterly attacked antagonist Stephen Jay Gould called Darwinian fundamentalism. Charles Darwin was not such a fundamentalist, as *The Descent of Man and Selection in Relation to Sex* makes clear. The *sexual selection* the title announces is a very different mechanism from natural selection, and many real evolutionary theorists consider it counter to natural selection. Sexual selection, and one of its most prominent generalizations, *signal selection,* can go against the idea of natural selection as a straightforward process, and can give rise to phenomena that can't be explained in the pure adaptationist terms that many evolutionary psychologists believe provide the only legitimate mode of explanation. Gould is anathema to them because he introduces the devil into the complex and irreproducible details of evolution. If many evolutionary phenomena are not reproducible, then there is no way to reasonably check any account of the origin of these phenomena—phenomena like storytelling.

The other possibility (offered by a different set of Gould's critics) is that storytelling is a purely accidental result of other human activities, and that not much can be inferred from the fact that we are so interested in stories that can't be inferred more directly by looking at the activities of which this interest is a side effect. I disagree with this claim as well. Understanding the *various* things that coalesce in storytelling and story-attending should affect our description of much wider cultural phenomena. To put it more modestly, perhaps, our practices of narrative can provide nuance as well as confirmatory evidence for certain more general accounts of how the modern human capacity for culture arose. I will show how some of these more general arguments are consistent with our storytelling practices.

My subject, therefore, is not the origin of storytelling but its psychological and biological conditions of possibility (to use Kant's term). What makes stories possible? What must we be like that storytelling is a human universal? (It is a human universal, but even if it weren't, it's certainly a widespread predilection among humans, and its explanation would cast light on the array of human capacities.) What is a story, and what must the mind have evolved to do to be able and eager to tell and hear stories?

In what follows I will give an account of the possible origin of fiction—of our absorbed and anxious interest in what happens to nonexistent beings. My argument relies on certain new ideas in evolutionary psychology and in the coevolution of species and culture. I will look at some recent accounts of the evolution of altruism and in particular of the oxymoronic form of altruism called *altruistic punishment* (punishment that costs the punisher more than it can net him or her). An innate capacity for and tendency toward altruistic punishment seems to me the central human psychological phenomenon in one aspect of our interest in narrative: our desire to see the good rewarded and the evil punished, whether they exist or not.

The model of evolution to which "Darwinian fundamentalists" subscribe is essentially Hobbesian, a genetic war of gene against gene. Organisms like us are servants of our genes, and we act to maximize their reproductive interest. In this model, we make alliances (as at a cellular level our genes do) to maximize our own genetic success. Behavior that seems genetically altruistic is, in fact, not so; we sacrifice ourselves for our genes, for our kin, for example, and according to degrees of kinship (Hamilton 1963; E. O. Wilson 1975), We also make implicit, genetically programmed behavioral contracts with others to take risks on one another's behalf if this improves our genetic chances overall (Trivers 1971). If we could get away with breaking those promises, we would; conversely, the more we can be seen to keep them the better off we are. This is the cynical view the narrator in Proust takes of M. de Norpois, of whom we learn that "the services he performed constituted not an alienation but a fructification of a part of his credit" (Proust 2.292).

On the Hobbesian view there is no such thing as true altruism. This thesis became dominant partly because no evolutionary accounts seemed able to explain how true altruism could evolve. But now such accounts have been given. Darwin himself had proposed a way for altruism to evolve through the mechanism of group selection. Groups with altruists do better *as a group* than groups without. But it was shown in the 1960s that, in fact, such groups would be too easily infiltrated or invaded by nonaltruists—that is, that group boundaries were too porous—to make group selection strong enough to overcome competition at the level of the individual or gene. More recently, however, a number of economists, biologists, sociologists, anthropologists, mathematicians, and evolutionary game theorists, working in a tradition derived from Adam Smith and especially David Hume, as against Hobbes, have argued for a pathway that the evolution of *genuine* altruism might have taken, and have concurrently demonstrated phenomena of genuine (nonkin) altruism, especially among humans. The pathfinders in this field are Robert Frank (1988), Sober and Wilson (1998), Ernst Fehr and his colleagues, and many of the people I cite below, most notably, for my early thinking about this, Geoffrey Miller. Miller puts their work together with that of the brilliant and delightful, and still controversial, naturalists Amotz and Avishag Zahavi, who have been writing about what they call "the handicap principle" since the mid-1970s, and who have been changing the minds of entrenched skeptics in greater and greater numbers. They summarize much of this work in their 1997 book. Although the Zahavis are skeptical of group selection (and therefore of genuine altruism), they give an extraordinary way to conceptualize the "evolution of cooperation" as Axelrod and Hamilton called it in 1981, and in particular of the evolution of cooperation among humans.[1]

These issues—altruism, cooperation, honest or costly signaling (which is what the handicap principle describes and explains)—seem to me central to an understanding of the human capacity for and delight in narrative. I'll begin by asking and trying to answer the questions: What makes us take an interest in fictional characters, that is, in other humans so genetically unrelated to us (see O'Gorman, Wilson, and Miller 2005) as not to exist? What would make us care about fiction? How do the general parameters required by the evolution of cooperation affect an answer to this question? I'll suggest a broad answer about why we might be interested in fiction, partly by way of an exposition of the theory of costly signaling or the handicap principle. (As a quick gloss: the peacock's tale is the classic costly signal.) We like costly signalers. There are interesting and relevant evolutionary reasons for this, which I'll rehearse.

Chapter 2 will explore signaling in greater depth, showing how signals work in everyday life, and in particular drawing a distinction between broadcast and focused signals. We can eavesdrop on broadcast signals, and here we can think about the ways that our interest in watching the way people signal to each other, talk and gesture to each other, might have evolved to the exquisite levels of sophistication that make naive absorption in drama possible.

Chapter 3 gives an account of what Philip Fisher (2002) calls "volunteered affect" and of the relation of the storyteller or purveyor of narrative, to its audience. Where Chapter 1 focuses on a more or less diachronic or vertical account of how an interest in stories might have evolved, and Chapter 2 offers a horizontal account of what particular narrative interactions we are peculiarly suited to overhear, Chapter 3 tries to draw this outward into a third dimension by describing in greater detail our own relations with characters and their purveyors. Here I treat the issues of signaling and cooperation that I argue are prominent within narrative with respect to the real event that narrative is (for example, campfire, religious drama, movie, TV).

In the last chapter I put the argument to work by explaining in somewhat more extended fashion what I take to be the powerful psychological grip of two exemplary literary scenes, one from *Oliver Twist* and one from *King Lear.* Everyone will recognize their power; I hope as well that everyone will recognize their own responses in my reading of them. That chapter will suggest that the issues of vindication and vindictiveness, broadly construed, capture pretty generally what we most care about in narratives. But in a coda I also very briefly suggest some other things that narratives do that appeal to the cooperative principle central to my argument.

My hope in this book, which is very different from earlier attempts to apply evolutionary psychology to literature, is to give an account not of what narrative should be but of what it is, and why it should be as strange, complex, and intellectual—as *cognitive*—as it is.

1

How Could an Interest in Fiction Have Evolved?

Interest and Learning

Why do we care about what happens in a fictional representation, to fictional characters, in a fictional world? What makes it possible for us to have a vivid emotional response to fictions? It's well established that an interest in narrative—in stories and storytelling—is a cultural universal (Brown 1991; Tooby and Cosmides 2001), but it's a puzzling one. I would like to suggest instead that certain considerations in recent evolutionary theory, especially around the "evolution of cooperation," can help to make sense of the puzzle of narrative interest, by which I mean anxiety on behalf of and about the motives, actions, and experiences of fictional characters.[1] This is a puzzle explicitly noted by Augustine, and probably implicit in Plato, who didn't like fiction. Aristotle, who did, tries to solve the puzzle by invoking the concept of imitation. My aim is to argue that certain central narrative motifs derive from and recruit capacities developed by natural selection for more fundamental reasons. (I hope this argument also provides evolutionary biologists with further evidence that such capacities are a feature of the human mind.) These capacities and the evolutionary problems they solve not only make narrative possible; they also make some kinds of narrative inevitable.[2]

Why do we willingly suspend disbelief, if we do? In Samuel Taylor Coleridge's famous, though usually simplified, formulation we suspend disbelief in the supernatural to keep faith with the representation of plausible humanity in implausible circumstances (*Biographia Literaria* 1969: Chapter 14). Coleridge is essentially looking back to Dr. Samuel Johnson's view of tragedy as appealing to how we *might* or *would* feel in certain extreme circumstances, and forward to A. D. Nuttall's concurring idea that we practice for the game of life by attending to tragedy (Nuttall 1996). I am interested in a less sophisticated phenomenon, though, that of actual anxiety or suspense or relief or gratification or rage or embarrassment or some other volunteered affect, to use Philip Fisher's helpful term (2002), on behalf of, or in response to what happens to, characters we know are fictional.

I would like to give an explanation of our responses to plot- and character-conveying practices (narrative, drama, film, dance, and opera as well, but here only implicitly) where we feel at least a partial and substantial portion of the strong emotional response to what could not be actually happening, that we would feel if it were actually happening. It could not be happening, in general, because if it were happening we would not be there to witness it, or to hear or read its narration. It's been a long time since anyone thought the unities were required to make dramatic illusion possible. Well, what does make dramatic illusion possible? What grip does the fictional have on us, and why?

By fictional I'll usually mean "nonactual," a term meant to apply both to purely fictional figures, who never underwent the incidents being narrated of them and who do not exist, and to figures who may, in fact, have experienced and had a hand in factual incidents truthfully narrated,[3] but who nevertheless are not doing so at the same moment that the audience—reader or spectator—feels anxiety on their behalf. It is tempting to say that anxiety about purely fictional beings derives naturally from the anxiety felt during the narration of true stories about real beings, since what counts is the story and not its actually taking place at the time of narration. We may feel suspense reading history, or hearing the true story of someone's narrow escape from disaster, or grief or Schadenfreude at their failure to escape, and fiction might be supposed to capitalize on this. But the very suspense we may feel about what is not actually happening suggests that it is more reasonable to reverse emphasis, and to say that an innate disposition to be interested in the narration of what is not actual makes possible the indispensable kind of learning that we attain in hearing truthful narratives. If such learning is genuinely indispensable, our instinct to learn this way must be a product of evolution.

Evolutionary psychologists have quite reasonably said that being able to learn through the experiences that others narrate is essential to human adaptation in a highly various and tricky world. Narration can efficiently show what a person or group of people have done and how and why they have done it, and can show that these things are causally linked. Furthermore, narratives are repositories of information, of knowledge that belongs to a group or population and that can be passed down the generations. Sugiyama (2001) gives an impressive version of this way of understanding narrative, making use of anthropological as well as biological evidence on the universality of narratives and the importance at least to Pleistocene populations of the subjects they treat and the characters they marshal. But she assumes without explanation that the interest, anxiety, and pleasure we take in *stories* follow naturally from the information they abstract and communicate. Like Samuel Richardson, she understands narrative as a delivery system for an important and salubrious core of precept and doctrine. She explains how narrative instructs, but not how it delights. Her theory, which I am willing to accept as far as it goes, does not give an account of fiction itself, that is, of why something that cuts against the notion of narrative as imparting accurate information, should have arisen. What happens when the poet nothing affirmeth? Why do we find some stories sweet, or relish others because they are "vinegared with satyres" (Mackenzie, 1660: 7–8)?

It is because narrative delights that fictional narrative can flourish. Evolutionary psychology has sought to explain fiction proper, stories frankly made up, as either a kind of happy by-product of truthful narrative (Pinker 1997) or indeed a further pedagogical adaptation, enabling individuals to understand and sympathize with the variety of human types and interests (Joseph Carroll 2004).[4]

I want to stress the way this kind of explanation makes use of a central tenet of evolutionary thinking, that evolution often proceeds as organisms make novel use of already established underlying capacities which they *recruit* to unanticipated uses. Fiction-making proper, as Carroll understands it, would recruit our capacity to learn from stories and extend it, thereby enabling a salutary and biologically successful extension of the range of human sympathy.

In what follows I'll tend to use *fiction* as the strongest test case for the argument I make about our interest in the nonactual. My account, however, is about the nonactual in general, not the purely fictive in particular, where by pure fiction one might more or less mean narratives that the audience, readers, or spectators know not to be happening, not to be true, and not to be relevant to themselves personally.

I think that Joseph Carroll (2004) may be right, in that fiction does have a capacity to extend the range of human sympathy. It's a claim consonant with Matthew Arnold's idea that literature spreads sweetness and light just as much as its natural exemplar, the honeybee. It is consonant as well with suspicious views of the regulative functions of literature, as in some followers of Foucault. But this explanation does not solve the basic question of why fiction should be able to keep us on the edge of our seats. Neither does Pinker's explanation. Human interest in narration may be the natural form of attention necessary to learning important facts about the world and to learning the causal relationships among them, and fiction may then trade on this interest. But a theory that stops there neglects the fact that narrative often relies on intense anxieties and gratifications counter to what accurate learning would encourage. The emotional intensity that a story will often elicit can make it very easy to believe in implausible solutions to the problems set. We are willing, and even grateful, to accept supernatural or magical or improbable answers, answers that are dysfunctional for learning how to cope with the real environments of our lives. (Even if someone has a taste for naturalism, this taste is notoriously not natural itself but cultivated.) The very fact that we're prone to accept and even demand magic solutions seems to indicate that the theory of narrative as efficient teaching can't be the whole explanation.

Another reason for listening to true narratives should also be noted: learning the truth about the behavior and therefore trustworthiness of various particular people we may have dealings with and must trust or reject. In James Merrill's biological language about the chimpanzee Miranda who represents the profoundest elements of primate life, our life ("Who can doubt she's one of us?"), the choice in our relations to others is either to "love, or overlook, or maul" (Merrill 1982: 18–19). For any person the decision to trust another will depend on that person's innate propensity and experience, and on the more or less obvious qualities in the candidate for trust. Is he lean and hungry or fat and jovial? Since there may be no art to find the mind's construction in the face, previous dealings and reputation will also be of great importance. Reputation will often be a matter of narrative, and since the people you may or may not trust can be of great emotional importance to you—a mate, a parent, a child—the truth or gossip that we learn about them may be emotionally gripping. The question whom to trust may be another important source of our anxiety about what narratives tell, an anxiety that fiction can recruit.

This argument might even account for our interest in narrative itself as opposed to the bottom line or ending. Narrative explains actions and allows for considerable subtlety in our judgment. As Austin insisted in "A Plea for Excuses" (1970), an essay too much assimilated to theories of performance at the expense of theories of narrative, people do things inadvertently or unintentionally or accidentally or unknowingly. Witnesses in courts of law narrate "in their own words" what they have witnessed, to make it possible to understand not only what happened but also why agents acted as they did, with what mitigation or aggravation of their actions. (Austin and the great legal theorist H. L. A. Hart worked through their categories together, and Austin thought the history of law an important source of the social, philosophical, and linguistic distinctions he was making, hence the word "plea" in his title.)[5] Conversely, as Elizabeth Loftus (1996) has shown, eyewitness testimony is peculiarly liable to be distorted by narrative suggestion that can override what we see "with our own eyes"; again, this suggests the central importance of narrative to our judgment of the behavior of others, since it can make us misperceive or misinterpret the actual testimony of our senses. (The blind Gloucester doesn't hear the sea in *King Lear,* because he and Edgar are nowhere near it, but Keats heard it. He describes in a letter how Edgar's deceptively suggestive line "Do you hear the sea?" haunted him until he wrote a sonnet about it.)

Nevertheless, this argument, even modified, is not enough. It still does not account for our strongly emotional interest in *fictions;* that is, characters whose fates we are not concerned in. It may benefit us to know whether we can trust someone, but it doesn't help to know whether we can trust Roger Thornhill or Eve Kendall (or Cary Grant and Eva Marie-Saint, for that matter). Fiction makes us care about characters and their fates as though we were anxious about real people of real importance to our lives. If narrative were primarily a vehicle for learning about the world around us, we would do a lot better learning its lessons dispassionately. Passionate interest—anxiety about unlikely dangers and gratitude for implausible solutions—is the least reliable mode of curiosity and therefore unlikely to have evolved as a way to facilitate the transfer of important information about the real world. Why then do we care so much about fictional or, more generally, about irrelevant characters?

One Freudian answer might be that they are not so irrelevant, because of what psychoanalysis calls the transference. We make of strangers the avatars of the most important people in our own lives. We transfer onto Roger Thornhill and Eve Kendall the anxieties we have about people who do concern us closely. The language of poetry naturally falls in with the language of power, as Hazlitt said, maybe because the language of power is the language of our parents. This insight and the terminology it invents do not solve the problem but reformulate it. Why do we engage in transference at all? Why do we extend it to fictions, to people in no way part of our lives? Why *do* we treat fictional characters like parents—or like siblings (as in Freud's *Totem and Taboo* [1913]) or children or friends or partners or lovers or rivals?

All the arguments that I have quickly rehearsed, even when formulated in the terms of evolutionary psychology, rely on a notion of narrative as mimesis that goes back to Aristotle and Plato. They all claim that we have an emotional response to fiction because fiction always represents or illustrates something else, something real: general human types; particular human beings whom each individual consumer of the fiction understands according to his or her own family romance; or essentially truthful lessons about how to judge and cope with the environment, including other people within it. In all these cases I might agree that fiction makes use of some of the capacities used in these other modes of human understanding, judgment, and action. But all these arguments share the assumption (sometimes without stating it) that it's natural for us to be interested in such representations per se. And this is the problem that we started with. We may be constituted to like reality represented through fictions, but why? Why do we care about what happens when it happens in representation, in fiction?

Imitation and Identification

I too also want to give an explanation based on some recent ideas in theoretical biology. I hope its elegance may recommend it, and may help with an understanding of how and why humans should have such a powerful emotional response to nonactual events and actions. And I acknowledge that looking to theoretical biology as I do is, indeed, to go back to the spirit and method of the first theoretical biologist, Aristotle.

I will begin by sketching an argument for why we are so good at following and taking an interest in nonactual events. In this argument I want to deny the tacit assumption that the audience's relation to the events that fiction presents or represents is that of the final consumer of fiction and the emotions that it can elicit. Evolutionary psychology has tended to a traditional understanding of audience response by seeing readers or spectators as "end-users" (which is why it tries to identify the adaptive benefits fiction brings to its audience). From Plato and Aristotle to Freud and Pinker, the spectator has been regarded as the final consumer of narrative.[6] The reasons we might wish to consume narrative vary, but Aristotle's therapeutic idea sorts well with Freud's, while Pinker's idea of pleasure in fiction as "mental cheesecake" (1997) is consistent with Plato's low valuation of its importance to human knowledge and human society. Joseph Carroll's critique of Pinker asserts that fiction can help us think about how to live, and help to humanize us, a claim conforming to Aristotle's idea that poetry is more philosophical than history. Both traditions agree that, for good or ill, fiction appeals to and satisfies the consumer's primordial pleasure in imitation.[7]

I'll assume that it is obviously true that fiction affords audiences pleasure through imitation. But I do not think that this is the whole story. I will argue from an evolutionary point of view that the aspects of mind that fiction appeals to have less to do with its potential therapeutic, malevolent, or harmless effects on the formation of our own minds than with its origin in and further development of a different capacity and propensity in the human mind. Rather than seeing our response to fiction as a kind of pleasurable and perhaps useful practice of judgment, with the judgment so practiced regarded as an end in itself, I will offer an account oriented toward human sociality and not pure judgment. (I mean judgment here in the Kantian sense of putting it all together in the mind. Kant's idea that aesthetics derives from the free-play of nonteleological judgment may be the culmination of an attitude toward fiction or discursive narrative that I am setting aside. But Kant was not talking about narrative.)

The framework that I will consider is that of *the evolution of cooperation* (Axelrod and Hamilton 1981). The evolution of cooperation is a major puzzle since it is hard to see how so surprising a phenomenon as human cooperation could arise through evolutionary processes. Cooperation is first a matter of relations between individuals. If the explanations recently proposed for the evolution of cooperation can shed light on narrative, what will turn out to be important is the audience's *relation* to the things imitated, to the nonactual events and actions recorded and represented, rather than the private pleasures and effects of the practice of imitation itself.

I will therefore argue against Aristotle's view of imitation as the most foundational category for drama. Aristotle saw imitation as an original propensity in humans: we learn by imitating, and take an interest in what others do in order to imitate them. Sometimes, too, they will help us to imitate actions that we are not yet competent to perform by imitating those actions themselves. We like to imitate and we like to observe imitation, the better to imitate; these are two conceptually distinct though related and overlapping attitudes toward imitation. The second one, our interest in seeing an imitation performed, is in normal circumstances an auxiliary of the first, but drama (and fiction in general) can isolate and recruit it to its own pleasure-giving purposes. This interest is for Aristotle (and, as suggested above, for most evolutionary psychologists as well) the root of our interest in drama, and perhaps of our interest in all narrative: we want to see or hear how a thing is done. Nature provides us with the pleasure we take in imitation as an *incentive* to learning. We take pleasure in hearing and seeing how it is done, and we are drawn to observe imitated action.

The idea of imitation as the foundation of all our response to drama has had misleading consequences. It is easy to see how such a theory led to the century-old standard (Freudian) idea of *identification,* by which we supposedly identify with the protagonist. Once the word takes hold people have a vague sense that identification means something like imagining ourselves in the protagonist's situation, even imagining that we are the protagonist. (Samuel Johnson already had a sharply defined sense of such imagination.)[8] Identification as it is now used bears the same relation to sympathy (similarity of feeling) as metaphor does to simile (similarity of some essential feature). But identification originally meant a kind of predication: something is identified as something else, as a kind of thing. Thus Hazlitt (1913) observes that it is the "abstractedness of our feelings in youth that (so to speak) identifies us with nature," that is *as* part of nature itself. Such an identification is made from the outside and is more like an objective judgment (in this case false) than an emotional commitment. Freud too thought of identification as a kind of objective assertion that the subject or self makes, treating itself as though it were an object that could be identified with (that is *as*) some other object or set of objects, while still remaining subjectively distinct.

But the popular psychoanalytic account of identification, which Freud sometimes encourages as well, does not see it as a categorical judgment (for example, that we belong to the category *nature*), but as an unquestioned emotional commitment rising to the level of metaphor or perhaps even the metamorphosis of audience into protagonist. Joseph Carroll (2004), who eschews psychoanalytic jargon, falls into talking about our identification with groups or clans we belong to, as though such identification had explanatory value. This version of identification essentially derives from the second aspect of imitation, whereby we imitate the action our guides have performed for us. No doubt such imitation is part of the penumbra of our response to performance—from playing the air guitar to using body English to try to keep a ball fair after the batter has pulled it down the line, to certain forms of sexual activity.[9] Kendall Walton's (1990) slogan—mimesis as make believe—captures the merging of the two ideas of imitation, and perhaps makes it possible to defend a less naive version of identification than the popularization of the psychoanalytic notion that has taken hold.[10] Richard Wollheim (1974) gives the best and most subtle summary of Freudian identification available, one to which I have nothing to object.[11]

Why am I skeptical of the popular use of the term in criticism, including the way Freud sometimes uses it? I do not think that the theory of identification does much to explain our interest in fictions. The widespread Freudian notion that we identify with literary characters (stated most succinctly and influentially in "Psychopathic Characters on the Stage" [Freud 1905]) seems to me at most to name rather than solve the problem of our anxiety or suspense about or intense interest in those characters.[12] I propose that vicarious interest is an irreducible and primary attitude that we take toward others.[13] Such an irreducibly vicarious attitude would go some way toward solving the problem about narrative presentation that I have already mentioned. The means of representation—storyteller, page, stage, or screen—logically ought to make it overwhelmingly obvious to us that the events we are concerned about are not happening. They are not like the dream events to which they are often assimilated, by Freud as well as by Kendall Walton (1990).[14] The storyteller or book, or theater, or projector, or screen is there where we are; the helpless hero in the dragon's grip is not. As Johnson insisted, we imagine ourselves neither in Rome nor in Alexandria.[15] And yet we feel anxiety for events supposed to be taking place in Rome and in Alexandria. Why do we forget the means of presentation or representation, and forget as well not only where we are but also in some sense *that* we are anywhere? If vicarious experience is irreducible, our own presence and our attitudes toward ourselves become a much less important feature of the situation. We forget the apparatus of representation as much as we forget the presence of our noses in our visual field. We are intent on what we are intent on, and not on how we come to be intent on it. We don't imagine ourselves in Rome or Alexandria because we don't imagine ourselves anywhere at all. Our imagination is directed elsewhere.[16] If we remembered ourselves, thought of our own position or perspective, we would want to know how we could know what we know. But, in fact, such a question doesn't tend to vex us, or to interfere with our concern or suspense, and this must be because our vicarious interest can be a primary and, paradoxically, a direct attitude.

All these considerations undermine the theory of naive identification that might be thought to solve them. If I imagined I *were* a character, I could not see her face; thus seeing her face means I must have a perspective on her that prevents perfect (naive) identification. In fact, as the movie *Lady in the Lake* (Robert Montgomery's 1947 adaptation of Raymond Chandler's novel) and some other disastrous experiments in radically subjective camera have shown by way of warning, attempts to encourage naive identification with heroes by showing us the world from their points of view only disorient and estrange us. We would rather see what someone else is doing than see the action through their eyes. When we try to see through their eyes it becomes vivid that their eyes are not ours, and we are brought to an intense awareness of the difference between us—the very opposite of identification. So identification can't in any straightforward way mean imagining yourself in the position of the person with whom you are identifying. As to identifying from the outside, I don't deny the possibility that a more sophisticated notion of identification would allow me some identification with what I see, but the sophistication weakens the explanatory power.[17] If I can identify with what I see, then my vicarious experience comes first, and identification is based on it. Such identification has nothing to do with some illusion that I *am* the character I care about.

I want to argue two consequences from the irreducibility of vicarious interest in humans: first, that an interest in others comes before any identification with those in whom we are interested; and, second, that such an interest will turn out to imply an interest in narrative, that is, in what others have done and suffered and in the causal relations among the things they have done and suffered.

For this reason I would like to hold on to the conceptual difference (one central to Freud as well) between the two aspects of imitation: liking to see it, liking to do it.[18] I wish to pose some further questions or make some further observations about the first of these aspects, that is about attitudes toward seeing an action imitated that are related less to the pure pleasure of imitation than to the pleasure of interested observation of what by its nature is often unobserved.

Conceptually prior to imitation is *tracking* or monitoring the agent or thing to be imitated. By tracking I mean perceiving an action or event. We humans are so good at learning because we're so good at imitating others, and we're good at imitating because we're good at tracking. It might be argued that imitation precedes tracking, and this is certainly true in individual instances. I look in the direction I see you looking, that is, I may imitate you by looking in the same direction, and then track what you are looking at. But I will do so because I perceive that is what you're doing. I imitate what I perceive you to do.

In fact, I imitate you only loosely, since my perspective on the thing you look at will be different from your perspective. If I were imitating exactly, and we were face to face, I would look west when you looked east. Or I would look behind you if I saw you looking at something behind me. But instead I look toward what you're looking at: I see *that* you're looking at something and try to see what it is. This requires what animal biologists call "gaze tracking." Humans have evolved very prominent whites of the eye probably to make our gazes strikingly trackable (a fact that silent film actors capitalized on). It's a fundamental fact about human learning, and the human interaction that makes it possible, that humans have to be very good at tracking one another's actions. I mean actions here in a vexed but undisputed philosophical sense: we don't just track physical correlates of actions—what philosophers call events. We also track what others are *doing:* looking, not just moving their eyes; pointing, not just lifting their fingers; spurning, not just pushing away; typing, not just depressing some keys (see also Gallese and Goldman 1998: 493). This requires what, using Premack and Woodruff's (1978) term, psychologists and anthropologists call a "theory of mind," a theory most children seem to develop no later than four years old and perhaps much earlier.[19]

I have two reasons for stressing the idea that tracking or monitoring is more fundamental than imitation. First, because this way we may be able to avoid the obscurities and misapplications of the notion of identification. Whether or not identification gives a satisfactory account of our relations to characters in fictions, it certainly doesn't give us a full account.

Second, because as a weaker notion tracking can capture more features and more elementary features of our response to fiction than imitation does.[20] Narration, for example, is not necessarily imitation, nor do we necessarily seek to imitate the agents whose adventures we follow. But we do follow their actions—using *follow* in a sense weaker than that of imitation.[21] We can recognize an action only by following it, and by definition agents are defined by their actions. It is evident that to follow a plot, in the most basic way, you have to keep track of what its agents are doing. (The standard present tense here—Hamlet stabs Polonius through the arras—is as reasonable a way as any to indicate the sense of immediacy that we feel in a gripping story.)[22]

How does it happen that the human mind has so pragmatically good a theory of mind, and is so successful in monitoring what other agents do or suffer? Fiction clearly relies on this capacity. What is it there for, and how does it affect our relation to fiction? The simplest answer to the first question is that being able to track what other people are doing and what they have done is necessary to any voluntary human commerce, especially with unrelated individuals whose success in life doesn't help one's own genes.[23] You have to know whom to trust, and why, and who trusts you, and why, and who has what skills or endowments, all of which refers itself to the others' past performances and their behavior in other cases of human interaction and commerce. One way that emotions function for us is that we respond positively or negatively, are attracted or repelled, by the way potential partners or rivals have interacted with other potential partners or rivals.[24] Human cooperation among unrelated individuals is a tricky business, and a surprising one from evolutionary first principles, and it has taken some ingenious experimentation and thought to come up with a workable hypothesis for how it has evolved. This hypothesis, which I will rehearse below, explains how evolution could have constituted us to have emotional responses to how others interact with one another. But to have these emotional responses we first have to be able to track how they interact with one another. The interactions that we're tracking occur in a context that includes how they have interacted with one another in the past, and we have to be interested in tracking these things as well.

Even a strongly Freudian theory of identification stresses the idea of tracking the figure we identify with from the outside. The stronger our identification with a character, the more it will be balanced by repudiation or disavowal, which requires us to see him or her from an external perspective. So In *Totem and Taboo* (Freud 1913) the figure we identify with is made into a scapegoat we disown and punish. We observe the character from a position of achieved exteriority. We monitor his or her fate while distancing ourselves from it.

For Freud, though, this disavowal is still one mode of a more fundamental identification. Freud's consistent interest, like Foucault's, is in the subjective experience of punishment from the point of view of the punished. This often entails describing the point of view of the punisher but only because of the fundamental identification that Freud (and to some extent Foucault) argues that the punisher has with the punished (and vice versa). I think, however, that a biological explanation of punishment might allow for a different way of thinking about our relations to those whose actions we track.

I want to stress that the considerations I shall advance are consonant with P. F. Strawson's important account of our "reactive attitudes" toward the behavior of others. I cite that account here at some length because I think it gives a powerful, correct, and ineliminable description of why we react as we do to others' behavior. The evolutionary psychology that describes and explains altruistic punishment stands or falls on whether it is consistent with the philosophical and moral psychological *facts or data* presented in a page like the following one from Strawson, a passage that (as will be seen in what follows) gets exactly right our attitudes toward fictional as well as real people, attributes that it is my goal to explain. Strawson is weighing the moral effect of our sense of a person's responsibility or guilt for some untoward thing on our emotional response to that person. Guilt implies an agent's membership in a moral community; indignation implies an interested observer's acknowledgment of the guilty agent's membership in a moral community:

The concepts we are concerned with are those of responsibility and guilt, qualified as "moral," on the one hand—together with that of membership of a moral community; of demand, indignation, disapprobation and condemnation, qualified as "moral," on the other hand—together with that of punishment. Indignation, disapprobation, like resentment, tend to inhibit or at least to limit our goodwill toward the object of these attitudes, tend to promote an at least partial and temporary withdrawal of goodwill; they do so in proportion as they are strong; and their strength is in general proportioned to what is felt to be the magnitude of the injury and to the degree to which the agent's will is identified with, or indifferent to, it. (These, of course, are not contingent connections.) But these attitudes of disapprobation and indignation are precisely the correlates of the moral demand in the case where the demand is felt to be disregarded. The making of the demand is the proneness to such attitudes. The holding of them does not, as the holding of objective attitudes does, involve as a part of itself viewing their object other than as a member of the moral community. The partial withdrawal of goodwill which these attitudes entail, the modification they entail of the general demand that another should, if possible, be spared suffering, is, rather, the consequence of continuing to view him as a member of the moral community; only as one who has offended against its demands. So the preparedness to acquiesce in that infliction of suffering on the offender which is an essential part of punishment is all of a piece with this whole range of attitudes of which I have been speaking. It is not only moral reactive attitudes toward the offender which are in question here. We must mention also the self-reactive attitudes of offenders themselves. Just as the other-reactive attitudes are associated with a readiness to acquiesce in the infliction of suffering on an offender, within the "institution" of punishment, so the self-reactive attitudes are associated with a readiness on the part of the offender to acquiesce in such infliction without developing the reactions (e.g. of resentment) which he would normally develop to the infliction of injury upon him; i.e. with a readiness, as we say, to accept punishment as "his due" or as "just."

I am not in the least suggesting that these readinesses to acquiesce, either on the part of the offender himself or on the part of others, are always or commonly accompanied or preceded by indignant boilings or remorseful pangs; only that we have here a continuum of attitudes and feelings to which these readinesses to acquiesce themselves belong. Nor am I in the least suggesting that it belongs to this continuum of attitudes that we should be ready to acquiesce in the infliction of injury

> on offenders in a fashion which we saw to be quite indiscriminate or in accordance with procedures which we knew to be wholly useless. On the contrary, savage or civilized, we have some belief in the utility of practices of condemnation and punishment. But the social utility of these practices, on which the optimist [about rational social engineering] lays such exclusive stress, is not what is now in question. What is in question is the pessimist's justified sense that to speak in terms of social utility alone is to leave out something vital in our conception of these practices. The vital thing can be restored by attending to that complicated web of attitudes and feelings which form an essential part of the moral life as we know it, and which are quite opposed to objectivity of attitude. Only by attending to this range of attitudes can we recover from the facts as we know them a sense of what we mean, i.e. of all we mean, when, speaking the language of morals, we speak of desert, responsibility, guilt, condemnation, and justice. But we do recover it from the facts as we know them. We do not have to go beyond them. (Strawson 2003: 90–91)

Punishment

While it is convenient for me to mention Freud or Foucault on punishment as I just did, my interest is independent of theirs. I want to think about punishment by applying some insights from theoretical biology that are relevant to the phylogenetic rather than ontogenetic sphere. (Ontogeny describes the development of the individual, phylogeny of the kind or species.) Freud more or less answers the question, "Why do we punish?" by analyzing why *I* punish. I want to answer the question, "Why do I punish?" by considering the answers given by evolutionary game theorists to the question, "Why do *we* punish?" What demands did Nature, that is, evolution, meet in giving us a disposition or instinct to punish?

I have said that our ability and desire to track others must be more fundamental than our propensity to imitate or to take pleasure in imitation pure and simple.[25] I will now argue that we have explicitly evolved the ability and desire to track others and *to learn their stories* precisely in order to punish the guilty (and somewhat secondarily to reward the virtuous); we have specifically evolved an innate tendency toward what evolutionary theorists call "strong reciprocity."[26] Strong reciprocity means that the strong reciprocator punishes and rewards others for their behavior toward any member of the social group, and not just or primarily for their individual interactions with the reciprocator. Strong reciprocators mind everybody's business, and in doing so, particularly when they punish those who cheat others, they insure social cohesion.

Their emotions are almost certainly the proximate causes for the reciprocators' reactions. Human emotions are, among perhaps many other things, mechanisms by which evolution insures that informed observers will have an immediate response to the treatment of others by still others: the response, for example, of grief, outrage, approval, satisfaction, vindication, scorn, relief, gratification, delight. (I'll return to the question of how an evolutionary mechanism could inculcate such capacities.)

Most humans seem to have an innate capacity to become strong reciprocators.[27] As with most human characteristics, strong reciprocity is variously and differentially realized depending on social structure, institutions, context, and individual experience, depending, that is to say, more or less on culture, including such cultural modes of communication and transmission as literature.[28] A general and systematic, though still speculative, account seems to be coming together about why we are so disposed to get involved in business that is of no direct concern to us. The most interesting and well-attested aspect of this propensity toward strong reciprocity appears in what biologists call "altruistic punishment."

To come to an account of altruistic punishment we first have to give an account of altruism. True altruism is by definition irrational, in the sense of rationality given by rational choice models of economic behavior: that is, I choose those things that reason would expect would yield the optimal outcome for me. An act of true altruism abjures the optimal outcome and is therefore irrational. To give an account of how true altruism might evolve we have to review a more overarching account of how and why irrational behavior in general, and altruistic behavior in particular, might yield better results for a person than do calculated and rational behavior.

To anticipate, I will argue the following points: First, irrational behavior can be a paradoxical advantage within a social system.[29] Second, altruism is a form of such behavior. And, finally, altruistic punishment is the form of altruism most necessary to the maintenance of cooperation within a society made up largely of nonkin, that is, most human societies.[30] I'll then turn to the question of how evolution might select for altruism in general and altruistic punishment in particular,[31] and what consequences this might have for understanding our relation to the non-actual and to fiction as a subcategory of the nonactual. To begin to do this we have to consider the game-theory set-up called the "prisoner's dilemma."

The Prisoner's Dilemma

True altruism—behavior that doesn't give preferential benefit to copies of one's own genes over those of unrelated members of the group—presents a puzzle for evolutionary theorists since their ultimate mechanism and arbiter is reproductive success. The puzzle is in the way that altruism is irrational, and therefore unlikely to evolve according to the highly rational selections of genes that nature makes, even though it can yield better results than rational calculation—if only others are irrational as well. The paradox is captured most succinctly in the famous game-theory conundrum called the Prisoner's Dilemma. The Prisoner's Dilemma is a two-player game in which the outcome will be better for *both* players if they both cooperate, rather than if they both defect, but will be better for either player if he or she defects, no matter what the other does. Although often described in accounts of how cooperation could evolve or fail, it is worthwhile pausing to analyze the dilemma here.[32]

Two prisoners are separately interrogated about a crime they have, in fact, committed together. If they both deny the crime, the state will be able to imprison them, but only for a year on a lesser charge. The state offers freedom to either prisoner if he or she confesses and the other doesn't. The one who doesn't confess will get twenty years. If they both confess, they'll both get five years. Each prisoner will be able to calculate four possible outcomes: 1) if I confess and my partner doesn't I'm free; whereas 2) if I don't confess and my partner doesn't I have to serve a year in jail; moreover, 3) if I confess and my partner does, I have to spend only five years in jail; and 4) if I don't confess and my partner does, I will spend twenty years in jail.

In situations 1 and 2 my partner doesn't confess. I would do better, to the tune of saving myself a year in prison, to prefer situation 1 and confess. In situations 3 and 4 my partner does confess, and I save myself fifteen years in prison by opting for situation 3 rather than 4, and confessing myself. By the rules of the interrogation I can't communicate with my partner, let alone influence his or her choice. All I can do is consider the two possibilities: that my partner confesses or doesn't. And in either case, no matter what my partner does, it's better for me to confess.

If my partner is rational he or she is likely to make the same calculation, so that situation 3 (we both confess) is the outcome that will arise out of our best strategic calculations, and we'll both do five years. If we'd both shut up (situation 2), however, we will do only one year each. So it's better for both of us if neither of us confesses. For the pair of us taken together there are three possible outcomes for the two of us. From best to worst these are: two years in jail in total (as in situation 2); ten years in jail in total (as in situation 3); and twenty years in jail in total (as in situations 1 and 4, which for the brace of us are equivalent). The best outcome for the group is therefore situation 2; the best outcome for any individual is the worst outcome for the other, namely situations 1 and 4; and the outcome we'll achieve if we each pursue self-interest rationally is worse than it would otherwise be, the ten years we serve together under situation 3.

The important point here, often insufficiently stressed in popularizations of the Prisoner's Dilemma, is that each prisoner's knowledge of the other prisoner's rationality—including the other prisoner's rational assessment of his or her partner's rationality—will reinforce the tendency to defect. If I know that you know that I am likely to defect, my incentive to do so is intensified. And you know this too. And I know you know this too. So the result is that our rationality works against us. My knowledge of your rationality amplifies my incentive to protect myself by defecting. And I know too that your knowledge that this is so amplifies your incentive to defect.

The dilemma, however, might be analyzed as having another crucial feature. Your situation exactly parallels my own. Therefore I might expect you to make the same calculations I do, and to arrive at the same decision. These calculations should include the expectation of parallel calculation argued in the last sentence, and therefore the further calculation included in *this* sentence. If we trust each other to analyze the Prisoner's Dilemma sufficiently carefully, we will see that the best outcome we can hope for from a combination of analysis and trust in the analytic abilities of the other is situation 2—one year each. Our only chance for a light sentence (one year and not five) depends on my taking my own impulses not to defect (since it's better for both if neither defects) as a kind of guarantor of your impulses not to defect, and a guarantor that you will find a parallel guarantor about me in those impulses.

And yet, whether or not I have confidence that you will come to the same conclusions and not defect, rationally I should defect. Either you will have defected or you won't. My wishful good faith will make no causal difference. It is, in fact, a wish that wishful good faith should make a difference for you in what you do. But if it does make such a difference, rationally I should still defect.[33] Only irrational guilt, superstition, or conscience will stop me from defecting, or stop you from doing the same. Not to defect would be an example of irrational altruism on my part.[34]

Since evolution is about competition, the mathematics of game theory tends to work well in modeling evolutionary processes. The rational strategies that enter into planning what to do in games like the Prisoner's Dilemma can be just as well be the results of natural selection, in which optimal behavior will tend to prevail over time. (This is how artificial learning works in computers and neural nets, and according to Gerald Edelman's [1987] theory of neural Darwinism, how the individual brain itself comes to be tuned to the world, which is how in humans the brain learns rationality.)

But the Prisoner's Dilemma shows that a person's rational behavior can be a liability, even in the long run. On the one hand, he or she might miss out on opportunities available to others who don't have a reputation for consulting only their own advantage.[35] On the other hand, they are susceptible of being taken advantage of in situations where their resistance would cost more than their acquiescence. They'll object, but they won't sue. This is indeed what Bullingbrook (now Henry IV) complains of at the beginning of *1 Henry IV*: "My blood hath been too cold and temperate, / Unapt to stir at these indignities / And you have found me—for accordingly / You tread upon my patience" (Shakespeare, *1 Henry IV* 2001: I.iii.1–4). Knowing his rationality, his rivals have no reason not to take rational advantage of him. Bullingbrook is the opposite of Richard III, whose irrationality is effective (at least for a while) at terrorizing opposition. Contrariwise, Hamlet complains that he lacks the gall that would make oppression sufficiently bitter to provoke a morally appropriate but still reckless response in him, instead of continuing his rational but ineffective considerations of the situation (thinking too curiously on the event).

By contrast, irrational behavior can logically lead to long-term gain. In Prisoner's Dilemma, if I thought you, as my partner, might act irrationally and not defect, I might not defect either. Guilt, superstition, or conscience might sway me, especially if I thought you would take the virtuous course—swayed by guilt, superstition, or conscience. I might use my loyalty as a wishful warrant, a superstitious fetish, for yours (for your superstition and wishfulness). Such irrational impulses can play the role of catalyzing mediators. We're still in parallel situations, but now we might have parallel irrational motives that can save us, so that we do one year each instead of five.[36]

If you knew that I had a particularly tender conscience you might trust me to act irrationally and be more willing to do so yourself; likewise, if I knew you had a particularly tender conscience I might think that you wouldn't pursue your main chance and I might allow conscience to outweigh pure interest. So a sense of the other person's willingness to act altruistically will help amplify my own tendencies toward altruism, if they exist. And if they exist in me, I might be willing to consider trusting that they exist in you, as well. Thus, my knowing something about people's dispositions to act irrationally can actually be to their advantage. True altruism can yield a better outcome in the long run.

The stability of such a situation depends on two things: that others do have a disposition to act irrationally, and that I know or believe that they do.[37] If someone gets irrationally angry at a slight, I am more likely to take special care not to slight him or her (or, if I am vindictive, to do the opposite).[38] If someone is irrationally careful not to abuse a trust, I can trust him or her. Not otherwise does Shakespeare's Richard II trust Gaunt to banish his own son. Richard knows that Gaunt can be trusted to act against his own interest, to avoid an accusation of partiality, when he renders judgment against his son. This is a calculation that Gaunt learns too late:

> Alas, I look'd where some of you should say,
> I was too strict to make mine own away.
> But you gave leave to my unwilling tongue
> Against myself to do myself this wrong.
>
> (Shakespeare, *Richard II* 2001: I.iii.243–246)

While one can repose trust in someone who places moral commitment above self-interest (in this case, the deepest interests of what biologists call "reproductive success") it's never right to trust a pure rationalist, as Gaunt has done. It might seem therefore that Richard's rational opportunism (which I may be alone in stressing) is the winning strategy. But his very rationality alienates many of his vassals, so that ultimately he loses to the seemingly less opportunistic Bullingbrook. (Bullingbrook fakes altruism very efficiently.) Gaunt is trustworthy in a way that Richard isn't, and in the long run Gaunt's decency leads to the kingship of his son and the downfall of his less-decent exploiter. Richard's rational selfishness finally works against him.

Similarly, it's no fun to play poker with a purely skilled rationalist, and such players tend to find fewer games. Their skill loses them opportunities to make money—unless they can fake irrationality, which is what hustlers do.[39] In fact, players known to be highly skilled will occasionally and ostentatiously play without looking at their cards, in a kind of artificial irrationality. The point of such behavior is to attract bets from players who see that they have a reasonable chance to win and will bet high. Since the skilled player gives up the advantage of knowing his or her cards, the player's skill doesn't scare others away as much, and he or she ends up getting better games and winning more money in the long run. The self-imposed handicap is a real one, and yet yields better results in the end.[40]

Hammett, who knew about such things, narrates a conversation between Sam Spade and Gutman in *The Maltese Falcon* (1999), in which they consider the merits of torture as a way to make Spade talk. Spade points out that, rationally, torture would be a mistake, because Gutman can't kill him if he wants the bird, and without the threat of death to back it up (narratives were more innocent then), there are rational limits to what Gutman can do. Gutman agrees but points out that things can go wrong, which means that it's not rational to rely entirely on Gutman's skill—his embodied rationality. Gutman is advertising limits to his own rational behavior, like a poker player not looking at his or her cards, because those limits actually represent an advantage in his negotiations with Spade.

Spade concedes the risks that Gutman's lack of full control might expose him to, and then suddenly flies into a towering rage, acting irrationally himself and getting what he wants. It turns out that this is partly an act—though only partly and Spade himself may not be sure how much of an act it is—an act that demonstrates the advantage that irrationality can confer. A person's recklessness makes people careful not to provoke him or her, and so impersonated recklessness can be a useful skill. But it's still recklessness that's impersonated: if true recklessness weren't the norm among the seemingly reckless no one would fear the seemingly reckless.

What will make it possible for you to rely on my irrationality, say in a situation in which cooperation will help us both, but defection will help the defector? There are several related factors that will allow you to do so. Evolutionary biologists have provided a number of theories of true altruism, but there are a couple that seem to be particularly illuminating and that also may be results of a single evolutionary dynamic.[41]

Altruism

In the Prisoner's Dilemma, the partners as a pair do better cooperating; the individual members of the pair do better defecting. Natural selection operates mainly on the level of individual reproductive success (that is, on the level of the individual's genetic heritage, a heritage for which close kin also serve as repositories). The Prisoner's Dilemma provides a widespread model for all sorts of environmental challenges, in which cheating or defecting would be a better strategy from the individual's point of view than would be trusting or cooperating. It's better for me to hide food, if I can get away with it, than to share it. If you don't share your food and I share mine, I'll tend to fall behind in the survival of the fittest game; if you do share your food and I don't share mine, I'll have a golden opportunity to take the lead in the game. These are the choices that natural selection, acting according to implacably rational rules, would seem to force on us.

Indeed, rational choice might be described as choice dictated by the implacably simple facts of competition for reproductive success. Therefore an individual's purely rational choice—the choice that that would seem to constrain a successful evolutionary paradigm—would be always to defect within Prisoner's Dilemma scenarios. The argument is only strengthened from a gene's eye point of view. Reproductive success, for a human and for a human gene, means multiplying copies of the genes in question, which seems to cut against true altruism—altruism not directed at close kin. But humans cooperate and like cooperating. How did this come to be?

The various attempts to solve this problem have led to interesting discoveries about the nature and extent of genuine human altruism. The phenomenon to be explained is why humans are genuinely altruistic, for it is well established that humans *are* genuinely altruistic. The definition of altruism here is not a psychological one. It's not an explanation of altruism to say that it gives the altruist pleasure ("Why else would anyone do anything?"). The question is why nature would make altruism pleasurable. Pleasure is an incentive to altruism, as it is to eating and to reproduction. How and why did such an incentive to genuine altruism arise?

I need to stress again that though genuine altruism solves the problem of the Prisoner's Dilemma, the very point of the dilemma is that individuals acting rationally can't solve the problem. Since acting rationally means acting in conformity with the demands of natural selection, there's no obvious way that natural selection could give rise to practices of genuine altruism.

To solve this problem, biologists have tried to deny that genuine altruism exists. The apparent altruism of skilled poker players who handicaps themselves by not looking at the cards is, in fact, a conscious selfish strategy designed to maximize the payoff. In the short term it may also help their rivals' payoff, at their own cost. I note in passing that the act of not looking at your own cards must be announced to others. They have to observe it. Part of what makes it an effective strategy is that the players *signal* the fact that they don't know what cards they hold—that is, they signal the apparently altruistic gesture. We will return below to the central role of signaling benefits in the theory of true altruism.

The most obvious case of something that looks like true altruism but isn't is altruism toward close kin. This is clearly indicated in the behavior of social insects, where members of a brood can all be very closely related, sometimes sharing 75 percent of their genetic material, a situation that cannot arise with nonincestuous human reproduction. The social insects are nearly as self-sacrificing in their behavior as are, say, cells in a single organism. We would no more expect an ant to refrain from sacrificing herself for the colony when occasion arises than we would expect cells in our skin to rebel against whatever dying off is necessary to heal a cut.

Sociobiologists have tried to generalize a similar claim. In J. B. S. Haldane's famous formulation: it's rational for me to give up my life to save two or more siblings (who are each repositories, on average, of half my genetic heritage) or eight or more first cousins.[42] This is rational from the point of view of my genes, which do not privilege their physical existence but the information they embody. They are therefore indifferent to their survival as individual bits of DNA. They are oriented instead to the survival and reproduction of as many replicas as possible. Genes are oriented toward the reproduction of their forms, not of their individual substances.[43]

From this perspective it is easy to claim that self-sacrifice on behalf of closely related kin is not altruism at all from the point of view of the genes. Rather, it is genetic selfishness. In fact, so deeply ingrained is our tendency to make sacrifices on behalf of kin that we don't need the incentive of pleasure to do so. Sacrifice on behalf of kin seems more a reflex than a pleasure. Love is not a feeling, as Wittgenstein (1967: 89e) said, but a deep commitment, something that can be "put to the test," where the test is painful.

But this gives rise to an interesting speculation: That the index of true altruism is the pleasure it offers as a psychological reward, that is as a mechanism to do what we would otherwise abjure as harmful to our own rational interests—that is, it's a "specious reward" (see Ainslie 1975). In the paradoxical world in which biological problems must be met with rewards for acting against one's own rational—that is, genetic—interests, the fact that kin altruism doesn't need the obvious and attractive incentive of pleasure means that the commitment to our kin is more fundamental, basic, and well integrated into the fiber of our biological behavior than relatively recent and harder to sustain forms of altruism, which require the bait of pleasure. So our question is: how did biology come to provide this bait?

The other major attempt to explain altruism away is to see it as reciprocal, whether because direct reciprocators outperform nonreciprocators, as argued by Trivers (1971), or more narrowly as part of a tit-for-tat strategy in solving repeated Prisoner's Dilemmas. Axelrod (1984) has shown that when the players who have to decide whether to cooperate or defect have a history with each other, the best strategy is to cooperate if the other prisoner cooperated last time, and to defect if the other prisoner defected last time. The cooperation that arises from this best strategy looks like a form of altruism, since you are aiding someone else at a cost to yourself (remember that in every iteration you do better if you defect, no matter what the other person does).[44]

There is no doubt that these accounts explain a lot of apparent human altruism, which turns out, from the gene's point of view, to conform to a rational calculation of self-interest. But they by no means cover the whole story.[45] A large number of observations and experiments in the last quarter of a century have demonstrated that true altruism—cooperation in cases where a strict attention to one's own prospects would be the optimal strategy—does indeed exist.

Altruistic Punishment

I'll rehearse two of these observations and experiments. One phenomenon is shown by what has come to be called the ultimatum game, originated by Güth in 1982.[46] This is the simplest form of a game that shows how altruistic concerns about fairness can trump material gains. In the ultimatum game, two strangers are put into a situation where they have a one-shot interaction. Because they are strangers, considerations of kinship don't enter anyone's calculations, and the interaction is one-shot so that a tit-for-tat strategy can't develop or be proposed. In this situation, the experimenter publicly gives one of the players a large sum of money—sometimes as much as three times his or her monthly income. The person who receives the money has to propose a split with the person who doesn't receive the money. The proposal is an ultimatum. The other person can accept the split or veto it. If he or she accepts it, the money is divided as proposed. If the person vetoes it, neither of the players gets anything.

It is irrational for the responder not to accept any proposed split from the proposer. The responder will always come out better by accepting than by vetoing. And yet people generally veto offers of less than 25 percent of the original sum. This means they are paying to punish. They are giving up a sure gain in order to punish the selfishness of the proposer. It's not that they require a 50-50 split. But they do require a split that conforms to some idea of fairness (as S. A. Frank [1995] puts it). They repudiate noncooperators. This fact belongs to a range of phenomena that have come to be known as altruistic punishment: paying or sacrificing what one has to punish someone one perceives as behaving unfairly. The altruism in altruistic punishment consists in the fact that it costs to punish (as it does in the ultimatum game), and the punisher's willingness to pay this cost may be an important part in enforcing norms of fairness. We'll return to this idea below.

What is particularly interesting about the experiment is that it works both ways. Most responders will generally reject a split where they receive less than 25 percent of the money, and most proposers, intuiting this, will propose a split where they offer considerably higher than 25 percent. It's not necessarily (or perhaps even likely) that the proposers do this out of altruism. Rather, they do it out of a reasonably good sense that the responders will not be motivated only by rationality. If you have a sense of human nature in which you will expect others irrationally to repudiate a clear gain, just because the gain is smaller than they might think fair, you will tend to offer them a larger share of the pie, and this means their irrationality works in their favor. More important, this shows a tendency in human nature that will lead to cooperation (here between proposer and responder).

In such experiments, cooperation is at best a zero sum game, so they isolate the role in maintaining cooperation played by people's expectations that other people will take considerations of fairness into account in their choices or moves. The goal of cooperation here isn't to achieve a better aggregate result than could otherwise be achieved. Rather, people cooperate because they have a generalized sense that fairness will trump optimal achievement. They have a good sense of the perspective of their fellow players, and in particular of their fellow players' demands for fairness.

Many variations of the ultimatum game have been tested, in many different cultural contexts, and the results are pretty stable.[47] Among those variations are versions where players can develop reputations as cooperating or defecting, or groups can develop reputations, without anyone knowing individually who is cooperating and defecting, and so on. In all these cases, a fundamental sense of fairness seems to militate against people's playing by the optimal strategy, where optimal strategy is defined as the best thing you can do when your competitor is doing the best thing he or she can do.

Fehr and Gächter (1998) did an experiment in a sequential Prisoner's Dilemma. The first player could choose to cooperate or defect and then the second would act in response to the first. In this two-move interaction, there was a strong positive correlation between the first player's trust or cooperation and the second player's response (Fehr and Henrich 2003: 58). In an extension of this game, the first player could pay to punish and reward the second player for what the second player had done. Since this would be the last move of the game, it is irrational for the first player to pay out any more money to the second. But people do punish and reward in these conditions. And more important, the second players *expected* to be punished or rewarded for their responses—they expected their partners to act irrationally. This expectation is not *calculated:* it is a deep and deeply social sense of someone else's emotional commitment to a sense of fairness.

Humans have an irreducibly intuitive and accurate sense of how other humans will respond to certain situations. This response is direct and not mediated. We are not Dupins, imagining or projecting ourselves into the place of the other to wonder how he or she would feel or what he or she would do. We *know* how others feel and what they would do; we don't work it out. Vicarious experience is an irreducible and fundamental feature of human sociability. It is as direct, in some ways more direct, as direct experience.[48]

The ultimatum game may be contrasted to the dictator game. In a dictator game, the proposer can dictate terms, without the receiver having a veto. Not surprisingly, if the receiver has no veto, the proposer will be far more selfish, just because he or she can. Nevertheless, even in dictator games the split is rarely 99-1, although it does tend to be far more one-sided than the 3-1 or 2-1 of the ultimatum game. The point of the contrast between the dictator game and the ultimatum game is that it shows that it's not the proposer who tends to be altruistic, but—surprisingly—the responder. The responder will pay, with no hope of reward, to punish rank unfairness, and it is *that* fact, or that expectation, that tempers the ultimatum that might be rationally made.

The second set of experiments tending to show the existence of genuine altruism, in particular how central an altruistic sense of fairness seems to be, present variations of the dictator game. In one such variation a third player is introduced and observes a dictator game. This third player can pay to punish or reward one or both of the other players; whatever he or she pays is amplified by the experimenter, so that every dollar paid deprives the dictator of $4 or gives his victim $4. Note that the third player gets nothing out of paying to reward or punish except the power or agency to do just that. It is highly irrational for this player to pay to reward or punish, but again considerations of fairness trump rational self-interest. People do pay, and pay a substantial amount, when they think that someone has been treated notably unfairly, or when they think someone has evinced marked generosity, to affect what they have observed. This is the quality that has come to be called "strong reciprocation." Furthermore, it's a quality (like that of the responder's likely veto) that we tend to know other people are likely to have, so that dictators will expect to have to meet the approval of strong reciprocators or suffer the consequences.

All these cases are interesting not only because they demonstrate a propensity for altruism, but also, and perhaps more important, because they demonstrate that people will expect such a propensity in one another. This begins to solve the prisoner's dilemma, since such an expected propensity might induce me to behave altruistically even as I expect you to do so.

These mutual expectations about how others will feel and respond ought not to be surprising: altruism is by its nature a kind of willing acknowledgment of the interests of another, of a sense of the rights another has to being treated fairly. That acknowledgment of the other's point of view cuts both ways: both proposer and responder are thinking about each other and what they think it would be appropriate to expect the other to do, not only in terms of self-interest but in terms of a decent respect to questions of fairness or cooperation.

Altruism therefore, by its nature, has a strong component of vicarious experience.[49] One variation on the ultimatum game shows this. If the responder is offered a cut, not by another agent or person, but by a roll of the dice, the responder will take a far lower percentage than he or she would from another person. The demand for fairness that a responder makes when playing the ultimatum game with another person is directed at the other person as person, as having human experience. People pay to punish unfairness. Indeed, in experimental situations where it costs to punish unfairness, the perceived unfairness is even greater and people are willing to pay to punish just because of the unfairness whereby they must pay to punish, because the fact that they have to pay to punish aggravates the sense of unfairness. (Shylock is all the more outraged that the money he spends looking to punish Jessica is wasted. Punishing her is costly, and this makes him more spiteful.) The fact that it costs to punish unfairness makes people perceive the unfairness as greater. When they can punish for free, they punish slightly *less*, despite—or rather because of—the fact that punishment isn't costly for them so that one might expect far higher levels of punishment. And people tend not to punish at all a randomized cut of the wealth, even if they think that the outcome is unfair to them. We punish people, not situations.[50]

Rejecting an ultimatum is an emotional and not a rational response. Unpacking the intention behind such a response can tell us something about the structure and function of emotion. By rejecting an ultimatum I am communicating something to you, even if I'll never see you again. I am communicating the intensity of my sense that your behavior is unfair. I want you to *know* that it's unfair, and I am willing to forego what would otherwise be a clear if unfairly small gain to make sure that you do know it. In fact, I want you to know just how much I am willing to pay. There is considerably more satisfaction in rejecting an 80-20 cut than in rejecting a 99-1 cut; part of the satisfaction is that I am paying a fair amount to teach you a lesson. And what is that lesson? It is a lesson precisely in how much *I* am willing to pay to teach it. You'd sacrifice more, and I less, if I rejected a 99-1 cut. Rationally, I punish less by rejecting the 80-20 cut. But I like the idea of spurning a greater rather than a lesser value, and the reason I must like it is that the more I spurn the more I put you in the wrong by showing just how unfair you are. You are so unfair that I put by all rational considerations to punish you. This is something I want you to know.

At this point it's important to recall that the satisfactions of altruism (including the vindictive satisfactions of altruistic punishment) don't undercut the altruism itself. Satisfaction in a losing act or disposition to act is rather the sign of altruism. A few years ago, newspaper columnist Maureen Dowd jeered at Bill Gates, Warren Buffet, Ted Turner, and others for wanting to attach their names to large charitable donations. True altruism, she said, would be anonymous: "The loveliest kind [of giving is] anonymous giving, which has gone out of fashion in an era when charity is to a social climber what a rope is to a mountain climber" (Dowd 1996). And it's clear that the givers do want something in return for altruism. Many of them are social climbers and want reputations as altruists. But what Dowd undervalues is the fact that it's a pretty good sign that you *are* an altruist if you are willing to pay a lot to get the reputation of being an altruist. The reputation is not—from the standpoint of maximizing wealth—worth anything like what it costs. Only an altruist would willingly bear those uncompensated expenses. And it's also the case that if charity is an effective mode of social climbing, the social order that boosts the altruists is itself not so evidently meretricious as she implies. (We'll return to this idea below, in the context of costly signaling.)[51] Pleasure in altruism doesn't mean that you're not an altruist. It almost certainly means that you are.

These considerations lead to an interesting insight: from a biological point of view *spite* is a form of altruism, more specifically of altruistic punishment.[52] Spite consists of paying to punish, more generally of accepting a loss in order to enforce loss on another being. Its first theorist, W. D. Hamilton (1970), noted that there was not much evidence of truly spiteful behavior in the natural world. (He was considering the more general view of spite: self-sacrifice for the sake of hurting others.) Spite does not seem to garner even indirect benefits worth its cost. Hamilton thought of spite as a kind of mirror image of altruism toward kin. A being disposed to sacrifice *on behalf of* kin might seem equally disposed to sacrifice *against* unrelated conspecifics competing against kin. But Hamilton pointed out that, in fact, the situations are not symmetrical, not negatives or mirror images of each other. In a large population, saving one close relative is worth far more than eliminating one random nonrelative. If I save my only two siblings I double the chances of my genes' propagation. But if I eliminate ten rivals in a population of a hundred, then I have increased my own genes' prospects only by 10 percent, and not by 100 percent. (This observation is a mirror image of the standard free-rider problem, of which the Prisoner's Dilemma is a special case. I save $2 by jumping the turnstile, which doesn't hurt anyone else as long as everyone else is paying. If everyone took the same attitude, however, then there would be no rides at all. Spite can't *help* anyone else but its propagation does hurt.) Even spectacular acts of spite—acts we are all too familiar with in the mode of suicide bombings—don't make statistical sense. The literary exception proves the rule: Samson's destruction of the Philistines, in Judges as in Milton, destroys a sufficient percentage of his enemies to give his kin a better chance than they otherwise might have. But contrast this with a possible reading of Iago (one I offer only by way of illustration): Iago seeks to disrupt the erotic lives of both Othello and Cassio; in interfering with their reproductive success he also destroys his own, so that his spiteful genes are not handed down, which means spite should die out in the long run, or at least turn out not to be genetically determined.

It seems pretty clear to biologists that the motives for spiteful behavior that Hamilton imagined and rejected don't play any significant role in nature. But the term *altruistic punishment* captures a motive different from the elimination of competition. One thing that it can imply is the way that people can act spitefully—indeed, I would say most often do act spitefully—to demonstrate how unfairly they've been treated. (This is not the central aspect of altruistic punishment that evolutionary biologists are interested in, but it helps to delineate the psychological phenomena requiring explanation, and helps with that explanation as well.) The expressive content of spiteful behavior will have at its core something like this: "Look how angry you've made me by the injustice with which you've treated me—angry enough for me to cause myself harm." The grand example of this kind of sulking is that of Homer's Achilles, remaining in his tents, going so far as to give up reputation for the glory he might garner in the Trojan War by staying out of the battle, and this despite the more or less explicit fact that Achilles' life is to be short, so that all wasted time is significant. Thus too does King Lear express his impotent rage against his daughters by storming out of Gloucester's house into the matching storm on the heath. So does Antony whip Thiddias, and then send him back to Caesar with the message: "Look thou say / He makes me angry with him . . . And at this time most easy 'tis to do't" (Shakespeare, *Antony and Cleopatra* 2001: III.xiii.145–149): his self-destructive anger means to communicate by announcing and displaying its self-destructiveness.

Such an expressive purpose is also the use to which Gaunt puts his vote to banish his son in *Richard II* (quoted above). That banishment really is an instance of altruistic punishment, one in which the motives can't be the ones Hamilton considered. This is because Gaunt banishes his son, not an unrelated conspecific, and does so in favor of more distantly related kin like Richard. Richard wins the encounter because he banishes an unrelated person in exchange (Mowbray). But Gaunt's banishment of Bullingbrook allows him to make much of the fact that he has been unfairly treated. He *has* been unfairly treated, and the privilege this fact affords him of complaint doesn't compensate for the unfairness. Put otherwise, he pays for his reputation as altruist by actually being an altruist.

This is not the only case where Shakespeare depicts the severe biological altruism of punishing one's own children. (The grand example of such spite is Medea, who gives up her own genetic progeny to punish Jason, who is not kin.) Lear may hope that Regan and Goneril are bastards—and therefore nonkin (as opposed to the bastard Edmund's kinship to his father, Gloucester)—but Cordelia clearly is not a bastard, and his line about the man who "makes his generation messes / To gorge his appetite" (Shakespeare, *King Lear* 2001: I.i.118–119) suggests how the sadistic pleasures of spite can offer an incentive to (genetically, genuinely) self-destructive behavior. In these cases, the spite is directed at the very thing sacrificed, and one need only think of the spurned daughters in Shakespeare to see its passionately expressive purposes for those who spurn them: Leonato ("Do not live, Hero"), Lear, Leontes, Titus, and Brabantio.[53] Nor are sons exempt in Shakespeare, from Aumerle in *Richard II* to Mammilius in *The Winter's Tale*, though they tend to have an easier time of it.

Perhaps the most naked example of spite in Shakespeare is Shylock. It is just because he seems so actuated by principles of rational (therefore, highly selfish) calculation that Shylock's spitefulness is so striking. He is willing to pay money to see his daughter dead. From this perspective, Shylock's great fault seems not to be usury but spite, which can mask itself as usury:

> I would my daughter were dead at my foot, and the jewels in her ear; would she were hearsed at my foot, and the ducats in her coffin! No news of them? Why, so: and I know not what's spent in the search. Why, thou—loss upon loss! The thief gone with so much, and so much to find the thief; and no satisfaction, no revenge. (Shakespeare, *Merchant of Venice* 2001: III.i.81–86)

He is willing to pay for the satisfaction of revenge (against his daughter, against those she connived with), so it's not the money he regrets but the thievery and rank unfairness of it all, an unfairness compounded by the fact that paying to punish doesn't succeed in punishing.

Some aspect of spite is Shylock's defining characteristic. Why otherwise demand a useless pound of flesh if Antonio fails, after offering a loan without interest? Noticing this makes possible an interesting observation about the end of Act IV. Portia and the Duke become aware that Shylock's spite might defeat their designs on him. Their own more or less altruistic punishment can go only so far. Shylock is forced to give up nearly all his wealth. His response to that is to express a bitter willingness to give up still more:

> *Shylock:* Nay, take my life and all, pardon not that,—
> You take my house, when you do take the prop
> That doth sustain my house: you take my life
> When you do take the means whereby I live.

Portia: What mercy can you render him, Antonio?
(Shakespeare, *Merchant of Venice* 2001: IV.i.371–375)

Portia uses the occasion to encourage Antonio to demonstrate the mercy that has been conspicuously absent in Shylock. But she does this in response to Shylock's warning that he doesn't consider the residue of his wealth worth accepting. He is on the verge of rejecting an ultimatum that rationally he should accept. Should he reject it, he could still demand Antonio's pound of flesh, and then his spite (as we should surely call it) or altruistic punishment would ensure that the Christians' triumphalism would backfire.

This example (as I've interpreted the dynamics of the scene) confirms an aspect of Robert Frank's (1988) commitment view of emotions: that the credible threat of irrational action—of spiteful behavior—can be an advantage. But it can be an advantage only if the threat is credible, that is, if people like Shylock actually do act spitefully. But we know he does act spitefully, even though he may try to justify his choices as yielding economic advantage. He prefers the pound of flesh—that is the punishment of Antonio—to any amount of money that he could get for saving Antonio's life. His answer to Salerio and Solanio, when they ask him what use the pound of flesh could possibly be, is very nearly a definition of altruistic punishment:

Salerio: Why I am sure that if he forfeit, thou wilt not take his flesh,—what's that good for?
Shylock: To bait fish withal,—if it will feed nothing else, it will feed my revenge; he hath disgrac'd me, and hind'red me half a million, laugh'd at my losses, mock'd at my gains, scorned my nation, thwarted my bargains, cooled my friends, heated mine enemies,—and what's his reason? I am a Jew.
(Shakespeare, *Merchant of Venice* 2001: III.i.47–54)

Later Shylock confirms this spate of emotion when he jeers at the guarded Antonio: he places the desire for revenge, and the desire to have Antonio upon the hip, the desire to be able to jeer at him, above any rational calculation of value. (Gratiano is a kind of double for him in this respect, there to show Christian spite without the hypocrisy of the other Christians.) The desire for revenge is itself a kind of spite (as elsewhere,

I am not using terms like "spite" moralistically), since it cannot undo the damage it punishes and even aggravates by punishing. This is the wildness of the justice at which it aims.[54]

The altruistic quality of spite (or the spiteful form of altruism) consists in its immense efforts at expression, its consuming desire to communicate successfully with the object of its hate. This is what jeering is about: it is intensely interested in the point of view of the object of contempt. This element of vindictiveness in spite is something that we will consider in some detail below. Here I want to note that Shylock's sense of punishment as expressive and communicative extends beyond a vicarious sense of his victim's experience, for Shylock's main reason for punishing Antonio is Jessica's elopement, with which Antonio had little to do (and for which Shylock does not blame Antonio). It is Salerio and Solanio, and all the Christians who connived at Jessica's escape whom Shylock spitefully intends to punish by killing Antonio: "You knew, none so well, none so well as you, of my daughter's flight" (Shakespeare, *Merchant of Venice* 2001: III.i.22–23). His spite considers not only the victim he can exercise it on, but also and perhaps as importantly the experience of those like Solanio and Salerio who love that victim and who would act altruistically (in a more benign sense) toward Antonio. The crucial point here is that Shylock is capable of tracking, and communicating across, networks of vicarious interest, and we are capable of tracking the way he tracks those networks. He knows what they knew.

Altruistic behavior need not be spiteful for us to regard it sometimes with caution and distaste, which we do when it goes awry. Altruists mind everybody's business, and we can list in the ranks of altruists such characters as Austen's Lady Russell and Mrs. Norris, Dickens's Mr. Grimwig (of whom more below), and the indefatigable Mrs. Pardiggle, about whom people tend to mutter something "about gentlefolks minding their own business, and not troubling their heads and muddying their shoes with coming to look at after other people's" (*Bleak House* 1977: 156).[55] This is a point that David Hume makes in identifying some surprising modes of altruism.

Hume (2006) first argued for the idea that vengeance or other overweening assertions of personal prerogative are altruistic acts. For Hume most passions are unselfish, since they are directed at aims that cannot be deduced from a prior axiom of self-love. In his characteristic role as bracing contrarian, Hume defends the generosity of most of our emotional impulses, and to do that he shows that even passions traditionally seen as negative—such as ambition for power, fame, or revenge—are not and cannot be, strictly speaking, selfish. Nature frames and constitutes us, he says, to pursue such things for their own sake, not for ours. In the appendix to the *Enquiry Concerning the Principles of Morals* (2006) on self-love, Hume distinguishes passions directly motivated by our own interests from those that cannot serve self-interest:

> There are mental passions by which we are impelled immediately to seek particular objects, such as fame or power, or vengeance without any regard to interest; and when these objects are attained a pleasing enjoyment ensues, as the consequence of our indulged affections. Nature must, by the internal frame and constitution of the mind, give an original propensity to fame, ere we can reap any pleasure from that acquisition, or pursue it from motives of self-love, and desire of happiness. If I have no vanity, I take no delight in praise: if I be void of ambition, power gives me no enjoyment: if I be not angry, the punishment of an adversary is totally indifferent to me. In all these cases there is a passion which points immediately to the object, and constitutes it our good or happiness; as there are other secondary passions which afterwards arise, and pursue it as a part of our happiness, when once it is constituted such by our original affections. Were there no appetite of any kind antecedent to self-love, that propensity could scarcely ever exert itself; because we should, in that case, have felt few and slender pains or pleasures, and have little misery or happiness to avoid or to pursue.
>
> Who sees not that vengeance, from the force alone of passion, may be so eagerly pursued, as to make us knowingly neglect every consideration of ease, interest, or safety; and, like some vindictive animals, infuse our very souls into the wounds we give an enemy? (pp. 94–95)

I want once again to stress a central insight here: nature frames our minds or appetites to certain goals, and analysis shows that pleasure is the incentive (or proximate cause) but not the reason (the goal or final cause) for which we do many of the things we do.[56] Hume's compatibalism about free will should be recalled here. Hume sees no contradiction between asserting that our wills are free and that what we do is determined by what we are (which we cannot control): "What is meant by liberty, when applied to voluntary actions? We cannot surely mean that actions have so little connexion with motives, inclinations, and circumstances, that one does not follow with a certain degree of uniformity from the other, and that one affords no inference by which we can conclude the existence of the other. For these are plain and acknowledged matters of fact. By liberty, then, we can only mean a power of acting or not acting, according to the determinations of the will; that is, if we choose to remain at rest, we may; if we choose to move, we also may" (*Enquiries* 1975: 73). We can speak of both will and natural disposition (as does Strawson), and there is no inconsistency in using both names for the same phenomenon. In modern parlance our motives and inclinations are determined by our genes in response to the appeal of circumstances that modify, regulate, and otherwise affect the modes of their expression. The important thing to remember is that what we want is not necessarily what our genes want; it's the way they get what they want, which is likely to be something else. But what we want is just what we want, here the vindictive pleasure for which we are eager. What is the content of that pleasure? What exactly does Shylock want when he wants to diet his revenge?

Hume's compatabilist insight into the mind's relation to its interests will be central to a nonreductive version of psychoanalysis, as when Freud says we want to be "happy in our lives" as opposed to saying, for example, that we are simply instantiations of our drives. Indeed, Hume's account is central to any psychology seeking to show not the obvious point that we act in conformity with the sum of our desires under the sum of our constraints, but the reason that we have the desires that we do, sometimes even in the face of the constraints that we meet. To show this one has to consider the nature of the desires independently, and not seek their meaning simply in their origins or determining factors, such as their utility to our genes. As I'll argue below, such utility won't be the only thing that constitutes our desires, even if considerations of narrow utility are what the genes would prefer: our genes have brought us to a place where their narrow interests can be overruled by other interests—by *our* interests, by the desires that they instilled in us but over which they lost the complete control they had, and with which they must therefore compromise. We and our genes are partners in a Prisoner's Dilemma too. If we and they act according to the narrow interests of our differing perspectives they do less well than if they compromise, so that the genes that in the end do best, that are therefore selected for, are not those that look only to their own rational selfish interests.[57]

They do have interests, however, and we also have to compromise with our genes, taking a willing pleasure in behavior that might be good for them—like risk taking—but bad for us individually. Again, the important point here is that our genes give us incentives (in the form of pleasure) to behave in certain ways that make us willing to behave in those ways without thinking about our genes, without rationally maximizing our own selfish interests. They have reasons for making us social and for making us take pleasure in that sociality, but their reasons belong to a virtuous circle: *because* our fellows are social our genes will do better to join with that sociality than to try to beat it. Our genetic makeup has to acknowledge and make us engage with our peers' disposition to be social and to engage with one another, and by acknowledging this parallel genetic makeup our own contributes to it. It is therefore the case that human motives can't be reduced to a simple manifestation of genetic motives, but neither can human motives be understood as indifferent to them. From the genes' point of view, human motives must be *treated* as an independent variable; the genes then produce incentives to motivate us to actions independent of strict calculation of genetic advantage.[58] Altruistic attitudes like vengeance, spite, or the vanity that often drives them are perfect examples of such cooperative compromises between the relatively autonomous emotional agents we have become and the genetic matrix that has had to concede this much autonomy to us. The compatibalism here is one between our experienced desire to seek the pleasures of revenge for the sake of revenge (for example) and the genetic determinant that not only allows us this pleasure but actually constitutes it as a pleasure. We do it because we like it. Our genes like it too, but for quite different reasons. They like it because it's what works in a population in which we like doing it. Liking it becomes a relatively independent experience, one that shows that we are not purely selfish, and that our genes could not survive if they were purely selfish. As Hume saw, we have incentives to altruism, and those incentives do not contradict the fact that it really is altruism but rather reinforce it.

Some recent studies have confirmed Hume's view of the way we are constituted to take pleasure from the neglect of every consideration of ease, interest, and safety. Although I tend to be skeptical of claims based on brain imaging, it is suggestive that the pleasure centers of the brain seem to be involved in altruistic punishment. Most interesting for me is evidence presented by Quervain and colleagues (2004) that the mere *anticipation* of punishing a defector yields pleasure.[59] In their experiment they discriminated between the pleasure afforded by seeing punishment occur and the pleasure afforded by anticipating the possibility of being able to punish, and showed that the incentive to altruistic punishment occurs in the mode of anticipation, not observation.[60] Such anticipation, I shall argue, is part of the anxious pleasure that certain narratives afford us. We look forward to a resolution in which self-satisfied defectors are punished.[61]

In addition, these results may show why the actual representation of punishment is not particularly satisfying. Punishment is a joy proposed, and one species of narrative proposes that joy. But once Shylock is punished he must leave the stage; the spectacle of his continued punishment would give no satisfaction. That's why the death penalty is said to be peculiarly unsatisfying for survivors, even those who say they support the death penalty. The criminal will go to his death as Timothy McVeigh did, ostentatiously unrepentant, where the point of punishment was to make him *feel* it, make him feel that he was wrong, make him regret and repent what he had done. Or if he does repent, we tend to begin to feel pity rather than vindictiveness. Pure unrepentant fearful remorse might be the optimal state we wish to perceive in the criminal. But such a state is so nearly a contradiction in terms—unrepentant remorse?—that we are very unlikely to ascribe it to anyone. Perhaps the motive to severe punishment of apostasy consists in the fact that the fear was real, but the repentance false.

The attempt to delineate a maximally effective fantasy of punishment might govern such stories of apostasy as Exodus. Pharaoh's repeated changes of heart allow Moses and the Israelites iterated scenes of vindication without that vindication's ever having the effect of attracting sympathy to the repentant Pharaoh. But even that change of heart is tricky, since God hardens Pharaoh's heart, and Martin Luther felt that the quality of his punishment was beyond human understanding.[62] I am reminded as well of the brilliant satirical headline in *The Onion:* "Hijackers Surprised to Find Themselves in Hell" (September 26, 2001). Their surprise is what we want: they thought they were right, and didn't even fear death. But they were wrong, and would come to know they were wrong beyond the grave. Yet fear and regret would humanize them too much: their *surprise* at being in hell is about the most our vindictiveness can consistently contemplate with gratification. I wish them to know, with the full human shock of knowing it, that they are monsters, but if they could know it and be shocked by it, that would mitigate the sense of their monstrosity that I wish to cherish. Surprise is about the optimal compromise between wanting them shocked and pained and not wanting them humanized.[63] Something similar to *The Onion* fantasy is to be found also in *Paradise Lost* when Satan awaits the applause of the rebel angels for his corruption of humanity, and is surprised to hear instead a "universal hiss" (Milton 1975: 10.508); the rebel angels too are surprised that what they meant as praise becomes shame: "Thus was th' applause they meant / Turn'd to exploding hiss, triumph to shame / Cast on themselves from their own mouths" (Milton 1975: 10.545–547). They are surprised at their own hissing, and this seems to be the limit of what we can coherently intend to communicate to them through vindictive punishment: the punished will continue to intend evil and won't therefore deserve our sympathy or forgiveness even as they perceive with surprise how condign their punishment is; here is the fulfillment of God's prophecy that evil will redound upon Satan's own rebellious head.

I think that this general incoherence in fantasies of vindication is the psychological correlate of the neurological discovery that punishment is altruistic and that it takes an anticipatory form. Most generally, altruistic punishment or costly punishment takes the form of anticipation in the present of vindication in the future for acts in the past. The combination of present and future tenses in our fantasies of vindication makes possible the otherwise inconsistent doubleness of our relation to the objects of rage and spite. *Now* they think they've won, but they have another think *coming*.[64] We anticipate of the person we will punish how much that punishment *will* undermine his or her *present* complacency. The pleasure of the intention to punish is a present tense attitude toward a future occurrence. This attitude explains a number of features of narrative, a claim I will justify in some of the analyses that follow (particularly of *King Lear* and *Oliver Twist*). I think it also suggests that the motive for first person narration may derive from the way such narrative affords the pleasure of a present tense attitude toward a past act—past in reality, but anticipated in the narrative, and so anticipated by the narrator as well as the audience.

The Evolution of Cooperation

I will put two more pieces into place to round out an account of the arguments I have found most convincing about how altruism, particularly altruistic punishment, could evolve to make cooperation among nonkin possible. How can the generalized Prisoner's Dilemma be solved? Why won't defectors outcompete cooperators?[65]

The account I will rehearse confirms some of the psychological analysis of the experience of punishing that I have found bolstered by Quervain and colleagues (2004), and it will also suggest the role played by some other aspects of punishment—aspects both of punishing and of observing punishment—in our experience of narrative interest and narrative suspense. The psychology may flow from the biology; the psychology may be a mechanism that nature evolves to get us to behave in certain ways. But it is the psychology that I am interested in, the psychology and what the biology can tell us about how we might have developed the psychological attitudes toward narrative that we have.[66]

Recent models of human cooperation have relied on a kind of reverse engineering that has given us a better sense of the structure of cooperation. Much evidence has shown that it's not possible to establish an evolutionarily stable system (ESS) if defectors go unpunished. Here's a brief sketch of the argument, from the point of view of the system or group, made by Robert Frank (1988) and considerably revised and refined by many others, notably Axelrod (1997), and also Sober and Wilson (1998). Evolution selects for the fittest individuals, where the ultimate definition of fitness is reproductive success. The individuals who are able to amass the most resources will have the most reproductive—hence evolutionary—success. As the Prisoner's Dilemma shows, unpunished cheating always yields a better outcome than cooperating. But the Prisoner's Dilemma also shows that pairs (groups) that cooperate will do better than pairs that don't cooperate. Groups with many cooperators will outperform groups with many defectors. The tempting idea of simple group selection (proposed by Darwin himself) has been pretty much exploded as a way to solve this problem. Under plausible conditions, individual competition for biological success overwhelms competition between groups. Just the advantage of outperforming other groups is not enough to make particular groups evolve to be cooperative. The individuals making up the groups have immediate interests promoting defection that are far larger than their interests in seeing their group thrive over the long run. Groups are not as well defined or as stable as individuals, because they are not *in*-dividual, their membership is porous, and defecting individuals will infiltrate cooperative groups.[67]

So what keeps individuals in line? Why do they cooperate? The immediate answer to the question is that cooperation will tend to thrive if groups punish defectors (those who free-ride on others' cooperation). In a social system in which a subset of the population punishes defectors, defection may no longer be advantageous, under the plausible assumption that punishment detracts more from the defector's prospects than defecting adds to them.[68] An ESS could be established if defection were punished, because being a defector would tend to be a losing strategy when competing with unpunished cooperators. If it is a losing strategy, then the number of defectors goes down, and as their numbers decrease they become easier to spot and cheaper to punish, so that (up to a point where we reach equilibrium) the diminution of defectors over time leads to further diminution.[69]

It's important to recognize that even cooperators might not necessarily punish noncooperators. We have to distinguish between at least three distinct types of agent in a population: cooperators, defectors, and punishers. The punishers (or, more generally, strong reciprocators) keep the defectors in line. This system is not enough, however, as Axelrod (1997) has shown. The punishers (who may or may not be a proper subset of the set of cooperators, and may or may not intersect with that set) do not do as well as the nonpunishing cooperators. Punishment exacts resources from the punisher. Thus evolution would select against the punishers who might otherwise stabilize the system.

This is because nonpunishing cooperators benefit from the highly helpful contributions of the punishers, in much the same way that the defectors had before benefited from the contributions of the cooperators. In the game-theory terms of economics, the defectors may be regarded as free-riders, capitalizing on the cooperation of others. The nonpunishing cooperators then become *second-order free-riders:*[70] they rely on the punishers to police the defectors, but don't themselves pay the costs of punishing, and so they outperform the punishers in evolutionary success. Evolution would therefore select against punishers and weed them out; without punishers the defectors would regain the upper hand, the cooperators would be weeded out, and we would achieve as the endpoint of evolution the Hobbesian war of all against all.

Axelrod (1997) showed through computer simulation how such instability would arise. Robert Frank and others meanwhile had come up with a way to explain how the inherent instability of a system that discriminates against both the honest and the punishers of the dishonest might stabilize itself. Frank (1988) shows how some combination of acquired reputation and what he calls "difficult to fake" signals could make ready scrutiny of potential partners relatively easy. Such scrutiny would be cheap and cooperators would outperform defectors by virtue of cooperating, if they could identify other cooperators relatively easily. A reputation for fairness, a reputation for demanding fairness even at one's own cost, would assure someone looking for a cooperator that he or she had found one.

Both defectors and cooperators look for cooperators (you always do better in a Prisoner's Dilemma if the other player cooperates, whether you do or not). Shakespeare's Richard II is a masterful defector, and masterful as well in identifying cooperators. Just as he has earlier put his rival Bullingbrook's fate in the hands of Bullingbrook's father, Gaunt, when he departs for Ireland he puts the affairs of his kingdom in the hands of his sternly disapproving uncle York: "For he is just and always loved us well" (Shakespeare, *Richard II* 2001: II.ii.224). His disapproval is a warrant for his honesty and loyalty (which is what Richard means by "loved us well"). Bullingbrook, motivated by both altruistic punishment and by ambition (not, as Hume says, a rational passion either), is also able to identify cooperators with precision (including York), but unlike Richard he tends to defect much less: from Exton, yes, but not from York, for example. And it is his great insight to identify Aumerle as a cooperator, just because Aumerle has plotted, against his own interests, to kill him.

I will stress here the idea that we can assess others' likelihood of cooperating, or defecting, essentially by keeping track of them—of their behavior, reputation, evident interaction with those around them. Bullingbrook trusts York because York had supported Richard.

In Robert Frank's analysis, refusal to cooperate with noncooperators is formally indistinguishable from punishment. Noncooperators do not attain as many of the good things available to cooperators. They do better when it comes to public goods, listening to National Public Radio without paying, for example. But because a general disposition to cooperate, and to be known to cooperate, yields better results overall, cooperators will tend to outnumber noncooperators by a margin wide enough to make tireless and vigilant scrutiny of others more or less unnecessary. This means that an ESS can arise with a mix of cooperators and noncooperators, in which cooperation is the norm and some limited violation of the norm a bearable cost of doing business for the cooperators.

Axelrod and others looked instead at what would happen if the question of punishment were made central. Since punishers would tend to do less well than second-order free-riders, Axelrod introduced a mechanism in which second-order punishers punish nonpunishers, that is, second-order free-riders. This modification yields impressive stability. Henrich and Boyd (2001) showed that by generalizing the infinite regression here (so that nth-order free-riders are corrected by nth-order punishers, even while those who don't punish become n+1th-order free-riders), arbitrarily weak social institutions could ensure cooperation. Their analysis is probably unnecessarily recursive (although it works fine for my argument); the crucial point, already in Axelrod, is that an ESS can be achieved when an individual is punished not only for defecting, but also for not punishing an observed defector.[71]

This is the schema I want to consider for a moment. The idea is this: if I observe you allowing someone else to get away with cheating a third person, I will be inclined to punish (condemn, berate, complain about, turn against) you for not punishing another defector. For such a situation to take hold, it must be the case that nth-order punishers are tracking with some sophistication a set of interactions that include original cooperators, original defectors, and observers of the defectors, with this set reiterated perhaps a number of times. I must observe that *you* observe someone else's defection. And if I observe as well that no one else has observed that defection, and that you have also observed this, I will be all the more inclined to see you as a second-order free-rider. If I observe that you have observed but failed to respond to the fact that they failed to punish a defection, I will see you as a third-order free-rider, and so forth.

We now begin to see why a capacity to track little, nameless, but remembered acts of strong reciprocity and altruistic punishment might have been a central evolutionary achievement. Strong reciprocity, again, is defined as paying to reward or (especially) to punish when the payment brings no fitness benefits to the person who pays (although, of course, it may yield a psychological reward).

Its relevance to narrative is this: Punishment is a present response to past actions. We punish after we track or otherwise become aware of those actions. The nonactual makes a vivid emotional demand in the present, in order to incite those who come to know of certain actions that have taken place in the past to reward or punish those actions. Punishment is a costly response to such actions, a cost born in the present for actions over and done with, and therefore may be (or in one-shot interactions will be) irrational, that is altruistic. It must be motivated therefore by some built-in propensity in us to punish, even though punishment is irrational. That propensity is to be found in strong emotional responses to blameworthy (and praiseworthy) actions. Evolution therefore endowed us with these emotional responses, these desires to make mad the guilty. But we wish to appall the free as well—that is (not perhaps altering Shakespeare's meaning), we do not wish to allow innocent second-order free-riders the luxury of ignoring crime and guilt. (This would motivate part of Hamlet's bitterness at the whole court—an idea I'll return to below.) Therefore we track not only the original actor whose actions we wish to see reciprocated, whether through reward or more likely punishment;[72] we track as well those who are in a position to track that actor, and we track as well those in a position to track those tracking the actor. Fanny Assingham, c'est nous.

We could order the basic dramatis personae of a narrative according to this hierarchy: we track the villain or antagonist, the hero or protagonist who goes after the villain he or she tracks, the erotic partner who tracks and admires the hero who goes after the villain, and the confidants of both hero, erotic partner, and even villain—the "window characters" as Dan Decker (1998) calls them—who track the various things that everyone else is doing and tracking. As Decker points out, every character can at any time be a window character, a character through whom *we* are able to track what other characters are doing.

Let me summarize my argument so far. Humans cooperate, and continue cooperating, because we monitor one another's cooperation vigilantly. To give us an incentive to monitor and ensure cooperation, nature endows us with a pleasing sense of outrage at defection and a concomitant sympathy for the victims of defection—an endowment demonstrated by the prevalence of strong reciprocators. We all monitor the behavior of others and often punish and reward in response to what we ourselves track or to what we learn about them. Emotional involvement is the proximal or efficient cause of our tendency to reward or (more likely and more intensely) to punish. Such rewarding and punishing is altruistic. Among the kinds of behavior that we monitor through tracking or through report, and that we have a tendency to punish or reward, is the way others monitor behavior through tracking or through report, and the way they manifest a tendency to punish and reward. We are thus emotionally involved in other people's strong reciprocation and, therefore, in the actions or behavior of those to whom they are responding. We measure their responses by the actions to which they respond. In this way, we are constituted to take an intense emotional interest in the nonactual (since past actions make claims on a present response) and in the actors who involve themselves in adjusting the outcomes of nonactual events through strong reciprocation, that is, through rewarding those who are good (like Telemachus, the goatherd, and the nurse in *The Odyssey* [Homer 1996]), and punishing the cheaters (like the suitors and the maids). We are fitted to track one another and to track as well how others monitor one another and what they do when they monitor one another. What we wish to track is past behavior, including past tracking of past behavior, in order to respond in the present to that behavior. Fiction recruits this central capacity in human social cognition for taking pleasure in responding to the nonactual. It gratifies the proximal or psychological aim of our interest in what some have done and how others have responded. That aim is the pleasure we take in strong reciprocation, particularly punishment, a pleasure useful in nonliterary contexts as an incentive to altruistic punishment and presumably evolved for that reason. That pleasure is one of anticipation, and we take pleasure in anticipating altruistic punishment—enough so that we demand of others that they be altruistic punishers, and that we anticipate what will happen in ways that underlie our interest in seeing how things will turn out, our desire to follow events until their resolution. Likewise, our interest at any moment of a narrative is less in what is happening than in what *will* happen in the light of what *has* already happened, and in whether what will happen is what *should* happen.

I will make the following schematic claim, one more or less in line with a Freudian observation about the psychology of narrative. Our narrative interest is always engaged by an altruistic figure as protagonist or victim or both. Often this figure will be an altruistic punisher. I cite more or less randomly: Moses; Achilles; Odysseus; Orestes; Oedipus; Antigone; Aeneas; Dante's Virgil and his Beatrice; any knight errant, perhaps, including Don Quixote; many Shakespearean heroes, including Benedick, Portia, Hamlet, Edgar, Paulina, Prospero, and so on; Milton's Satan, Christ, and Samson; Emma Woodhouse; Daniel Deronda; Ahab and also Ishmael; almost any James heroine, particularly Catherine Sloper, the Governess, Isabel Archer, Milly Theale, Maggie Verver, and, to some extent, Lambert

Strether; almost any modern detective; and almost any modern superhero. We instinctively approve of what altruistic punishers do. We are constituted to observe what they do with some attention.

But how did it come about that we like *them?* What makes them heroes to us? This is the question that Freud tackles in *Totem and Taboo,* and it's a good one.

Costly Signaling

> If it be love indeed, then say how much.
>
> We are busy every moment giving our life its form, but we do so copying, despite ourselves, as though from a design, the features of the person we are and not of the person we would find it agreeable to be.
>
> —Proust 1987 2: 485

One more concept must be introduced here in order to explain what it is we admire about altruists in general and altruistic punishers in particular. The argument I'll rehearse here claims that the only way true altruism—strong reciprocity—*could* develop was by being admirable.[73]

It has long been recognized in evolutionary theory that certain characteristics are examples of conspicuous consumption. The peacock's tail is perhaps the most famous one: Darwin (in *The Descent of Man* 2004: Chapter 14) used it to explain what he called female sexual selection. Because peahens preferred glorious tails they were more likely to mate with peacocks with such tails, and their male offspring then also had such tails. Darwin saw this as an unconscious process of attraction to the beautiful, but another mechanism is more likely: the peacock's tail is a *signal* or index of his general fitness, and therefore of his value in providing a matching gamete to form a zygote in partnership or cooperation with the gamete containing half the peahen's genes. But the system is also interestingly self-amplifying. Part of a peacock's fitness is his attractiveness to hens. His tail is a sign of fitness (in ways that I'll explore below), but also an element of fitness because of its value as a sign. A hen's attraction to a well-betailed peacock is amplified by her (genes') interest in bearing male offspring with attractive tails, so that her genes can be passed on to more grandchildren.[74] Intergenerationally speaking, genetic desire is certainly mediated desire: parents desire in mates qualities that they wish their offspring to inherit in order to be attractive to potential mates in their turn. Notice that this too is an irreducibly vicarious attitude, a genetic analogy to the analysis of the Prisoner's Dilemma in which we trust that the other player will be just like us in trusting that we will be just like him or her. Vicarious expectations can be self-fulfilling, and through such self-fulfillment self-regulating norms may be instituted.[75]

Nevertheless, as R. A. Fisher (1915) was the first to point out, the signal cannot take on a completely independent life of its own. What appeals to sexual selection, even runaway sexual selection, still must conform to the demands of natural selection—of survival. Two imperatives must establish an equilibrium: to survive long enough to reproduce, and to be selected as a reproductive partner. But these are not entirely independent demands. The best reason to choose an individual as a reproductive partner is the evidence that the individual gives of being likely to pass along survival as well as reproductive probabilities to the chooser's offspring. After all, the offspring must survive long enough to reproduce. The decision concerning with whom to mate must be recursive, since reproductive probabilities are always a function of the next iteration's (the next generation's) survival and reproductive probabilities.

Erving Goffman (1981) notes that a move in a game does two things simultaneously. It announces itself *as* a move, and it does something within the strategic parameters of the game itself. So too does the kind of signal that the peacock's tail exemplifies: it signals fitness and it signals the fact *that* it's signaling fitness. It belongs to the class of biological phenomena that have come to be called costly or honest or difficult-to-fake signals.

Phenomena such as the peacock's tail are good signals of fitness because they are costly signals. Costly signals indicate that the signalers can bear the costs of signaling. Such signals have long been studied in human societies in anthropological considerations of gift-giving, especially in the extreme form by which prestige is derived from the conspicuous destruction of wealth, such as in the ceremonies of the potlatch among the Kwaikutl and Tlingit Indians of the Pacific Northwest.[76] Biologists have noted similar patterns of signaling in very many species, especially among vertebrates.

Amotz and Avishag Zahavi have been studying this phenomenon, which they have dubbed "the handicap principle" (1997), in great detail for more than three decades. Their idea is a very deep one, and in many ways highly surprising. What they proposed, and what Grafen (1990) showed mathematically, is a consistent theory of honest signaling, whereby such signaling works to the advantage of both signaler and receiver. The handicap principle, as they argue it, responds to a "need for reliability [that] links the signal directly to the message it conveys and guarantees that the cost of the signal is reasonable for an honest signaler but prohibitive for a cheater" (Grafen 1990a: 534). Receivers will evolve to learn what signals are reliable, based on the fact that they are costly and that they signal the ability to meet their costs, and signalers will evolve to put whatever energy they can afford into honest signaling.

They will do so because, as Grafen demonstrates, it's always the best strategy to signal your fitness as honestly as you can. Advertising a level of fitness greater than you have, when advertising is costly, is not worth the signaling benefit; advertising a level of fitness less than you have is also a losing strategy, because you don't derive the benefits due to fitness.

Grafen and Amotz and Avishag Zahavi make breathtaking claims about the possible domains of signaling, including, most strikingly, cooperative signaling between prey and predator. The idea behind costly signaling is that all honest signals should be wasteful. The message conveyed by waste of a resource is that the individual can afford to waste that resource. In Samuel Johnson's (Boswell 1998) acerbic formulation, "All censure of a man's self is oblique praise. It is in order to shew how much he can spare" (p. 972). It is such obliquity that is interesting in the theory of costly signaling as honest handicap. Johnson's asperity should be tempered with Hume's observation that self-praise, oblique or not, is an unselfish passion since it aims at communication with others. Thus do Amotz and Avishag Zahavi and Grafen see wasteful behavior as a kind of cooperative altruism, even among enemies.[77] Grafen describes mathematical "models that can be applied to many kinds of signaling . . . not restricted to the female choice type of sexual selection . . . It is theoretically coherent and consistent to say that roaring is used in fighting because it is energetically expensive, that nestlings beg so noisily because it reduces their growth, and that antelope stot because it reduces the speed at which they escape from lions. These apparently paradoxical ideas work because signaling systems require waste to ensure honesty" (Grafen 1990a: 527).

I want to stress again that these models show how cooperation *could* lead to profound ESS within whole ecologies and not only in populations of conspecifics. Although he doesn't cite him explicitly, in the passage just quoted Grafen expects his reader to see that he is contradicting a famous remark of Richard Dawkins (2006; first published in 1976) about the prevalence of *dishonesty* in signaling (Dawkins himself is summarizing Trivers, who wrote the introduction to the first edition of his book). Of nestlings Dawkins says:

> Many birds are fed in the nest by parents. They all gape and scream, and the parent drops a worm or other morsel in the open mouth of one of them. The loudness with which each baby screams is, ideally, proportional to how hungry he is. Therefore if the parent always gives food to the loudest screamer, they should all tend to get their fair share, since when one has had enough he will not scream so loudly. At least this is what would happen in the best of all possible worlds, if individuals do not cheat. But in the light of our selfish gene concept we must expect that individuals *will* cheat, *will* tell lies about how hungry they are. (pp. 129–130)

Where Dawkins sees the loud nestlings cheating the others, Grafen sees them as cooperating with their parents, to whom they honestly signal their fitness. The loudness with which each baby screams may signal how much each baby is thriving, not how hungry it is. The weak birds must conserve energy whereas the stronger can waste it by conspicuous complaint. To those that are likely to thrive more shall be given, and from those likely to die shall be taken away even that which they have. Again, it's worth recalling that this need not be—probably is not—the proximate psychological motive for the mother bird. But the psychological motive is dictated by a genetic disposition to cooperate with the bearers of her genes most likely to survive. Similarly, roaring stags indicate to rivals how much surplus strength they have and are willing to dissipate ("Our overplus of shipping shall we burn," says Shakespeare's Antony), which may spare them the actual and potentially greater costs of fighting (Antony turns out to be overconfident); and stotting antelope signal to lions that they are fit enough and fast enough to make hunting them less worthwhile than hunting antelopes that stot less, sparing both the stotting antelope and the hunting lion the costs of a difficult escape or a difficult kill.[78]

We can therefore see that certain kinds of public—trackable, monitorable, reputation-enhancing—expenditure can arise as a costly signal of fitness. With this general principle in mind, we can consider how altruism in general and altruistic punishment or strong reciprocity in particular might evolve as a costly signal or honest handicap—how it might develop through the dynamics of sexual selection. (Antony's magnanimity is what most impresses Cleopatra: "realms and islands were / As plates dropp'd from his pocket" [Shakespeare, *Antony and Cleopatra* 2001: V.ii.90–91].)[79]

A number of recent studies have suggested how genuine altruism or strong reciprocity can take hold within a population. The details of the account are fascinating, but here I'll rehearse only its broad contours. The crucial idea is that costly signaling is the more basic phenomenon. As Geoffrey Miller (2000) argues, the competition for reproductive success is the one that matters most from an evolutionary point of view. You need survive only long enough to reproduce, while long life without reproduction is an evolutionary dead end. Costly signaling, especially in situations where people have a great deal of choice in mate selection, is the best way to advertise yourself as a fit cooperator in the task of sending genes down the generations. Organisms have therefore evolved an enormous variety of costly signals.

Remember that costly signals eventually are constrained by the inflexible need to survive. A peacock that can't bend over to eat because its tail is too heavy, or that attracts predators because it is too conspicuous, won't survive long enough to reproduce. The signal of fitness can overwhelm the fitness it signals, not because of the waste it requires but because the signal itself—the hypertrophied attribute—detracts from fitness. If a muscle-bound person may lose a fight, it won't be due to the energy he or she has put into working out, but the loss of flexibility and agility that comes *with* being muscle-bound. Conversely, that detraction from fitness is part of what the signal shows its bearer can afford—but only to the extent that the bearer can actually afford it.

Some costly signals can produce secondary gains, however, in addition to the gains at which the signals aim. Rapunzel's long hair could serve as a ladder. Knowing how to dance might help in a fight. Being able to run a marathon is a better costly signal of the capacity for extreme bodily discipline than being anorexic. Achilles' acknowledged fatedness to a short life means that his sulking in his tents is a costly signal with no secondary gain for the Achaians. He purchases status through waste of whatever life remains to him, but the waste does no one any good. Achilles' battle with Hector, however, is a costly signal that also puts him at risk of immediate death (of the immediate fulfillment of his fate), but here the waste of a potential future aids all the Achaians.

Geoffrey Miller (2000) synthesizes a number of arguments about stone-age practices, particularly in hunting large game. Hunting such game is a cooperative enterprise, but the enterprise itself is not particularly efficient. It is far more productive to go after small game, where the hunting is less dangerous and more reliable. Large game is much harder to hunt, and even when the hunt is successful, the meat goes bad quickly.

It also appears to have been and to be standard practice among big-game hunters (or hunters of game with low net-yield in calories and protein) to distribute the meat that they do bring back equitably among kin and nonkin, among fellow hunters and among those unable to hunt (the feeble, infirm, sick, old, young), again without regard to kinship. These two facts lead to the reasonable speculation that big-game hunting among Paleolithic people no less than among Hemingway characters is a form of costly signaling.[80] Successful hunting already advertises fitness, and equitable distribution of the meat makes the signal more costly, thereby advertising further fitness.[81]

Now there are many other ways to advertise fitness, but successful hunting also contributes an important secondary gain: the protein the hunters distribute contributes to the success of all members of the population. If this population is competing for resources with another in which different costly signals are used—say jumping from high branches or fasting for days—the fitter population will outcompete the less fit one. Groups in which such behavior is the norm for displaying a handicap, and groups in which valuing such behavior is the norm for rewarding the fitness the handicap displays, will do better. As Fehr and Henrich write, "Within-group selection creates evolutionary pressures against strong reciprocity because strong reciprocators engage in individually costly behaviors that benefit the whole group. In contrast, between-group selection favors strong reciprocity because groups with disproportionately many strong reciprocators are better able to survive" (Fehr and Henrich 2003: 28–29). If there is a lot of between-group competition, then those groups whose modes of costly signaling take the form of strong reciprocity, especially altruistic punishment, will outcompete those whose modes yield less secondary gain, especially less secondary gain for the group as a whole.[82]

Gintis (2000) shows how under plausible assumptions a small number of altruistic punishers can keep would-be defectors in line. This is an important fact under conditions where calamities seem to strike groups on average every few decades. Under calamitous situations, most people will reckon their own chances as better if they save their own skin, but the group will do better if it cooperates. A small number of strong reciprocators can keep the group together under these conditions, and the groups that are kept together will insure that the genes of the strong reciprocators survive as well, especially if being a strong reciprocator confers status.

It seems intuitively appealing as an explanation of the kinds of heroes narrative tends to offer us to see those heroes as strong reciprocators—as cooperators, altruists, and altruistic punishers.[83] In particular, there is an eminently plausible way of construing the relationships between costs and values, as Gintis, Smith, and Bowles show, in which "there will be an equilibrium in which high-quality individuals will punish and low-quality ones will not" (Gintis, Smith, and Bowles 2001: 108). We admire such figures because we descend from populations in which altruistic signals of fitness throve. They still thrive. We admire altruists.[84] Knowing who is an altruist requires us to keep track of altruists. Knowing who is an altruistic punisher requires us to keep track not only of the punisher but also of those he or she has punished—and this will require us to juxtapose an account of the past or pluperfect with the altruistic punisher's later response to that account.

My own armchair speculation is that it's likely that narratives of altruistic punishment and hunting go together in stories of the hunt. The archeological evidence shows that modern humans learned far better, more difficult, more cooperative techniques of hunting in the transition from the middle to the upper Paleolithic period. This transition occurs at the same time that we start producing totemic and anthropomorphic representations of animals—cave art, sculptures, and so on. Humans will sometimes be buried with animals in what appear to be religious rituals. Steven Mithen (1999) argues that sophisticated hunting, the sophisticated tool-making that makes it possible, sophisticated archives of cultural skill and cultural memory, the art that constitutes those archives, and the religion from which it may not differ, all explode at about the same time, as various earlier cognitive capacities become integrated with one another. I am particularly interested in his suggestion that one of the most important innovations of modern human hunters was that of anthropomorphizing animals: we could predict what they would do (and therefore hunt them better) when we began seeing them not as rocks or stones or trees but as near human in their habits, wiliness, motives, and so on. We ascribe intentions to them, for the first time, and keep track of them and their intentions.[85] Totemic animals represent such anthropomorphisms, which led to religion (and no doubt to the punishing activities of angry gods), perhaps originally as a by-product, perhaps as a mechanism for anthropomorphism. I would like to note the extent to which this would give rise to the idea of hunting as a kind of altruistic punishment as well. Humans go after predators—the tiger that has decimated a village, for example. But even when going after prey, our tendency is to treat the prey as culpable. Think of Melville's Ahab, or of Françoise, Proust's more domestic Ahab, cursing the chickens who try to escape her slaughter. Or think of the more-or-less ritually slaughtered bull, the villain whom the heroic matador will bring to heel. It seems likely that narratives of altruistic punishment are natural and easy generalizations of narratives of hunting, and conversely that narratives of hunting were in their way as exciting as anything since, because they are also narratives of altruistic punishment.[86]

I should stress again that altruism as a costly signal or honest handicap is entitled to be called *genuine* altruism. Reviewing the order of events that plausibly lead to a genetic niche for strong reciprocators might help make this clear. Groups with genuine altruists outcompete groups without them, especially at times of great stress, so that a plausible process of group selection selects for those groups that contain altruists. The signaling benefits of *being* an altruist are not so great as the signaling benefits of being an amasser of wealth, since being an altruist is more costly than being rich. Not that being rich is cost-free—being a miser is very costly indeed; this is why wealth, especially the display of wealth, is a genuinely costly signal as well. But in a group in which those selecting mates esteem them on the basis of a signal that is truly altruistic rather than accumulative, true altruism will thrive, and *then* that group will thrive. Note that the selectors in such a group, as well as the selectees, will be more willing to pay the costs of such selection than the more rational selectors and selectees of the group that signals through accumulation of wealth rather than altruistic distribution of wealth.[87] The true altruists depend on the true altruism of those randomly arising people who prefer true altruists to rational maximizers. Altruism isn't a covert form of selfishness, but relies on the altruism of others, on altruistic approval of or general benevolence toward altruists.[88] An altruistic mutation casts its bread upon the waters, and in certain circumstances finds it again after many days when unexpected, unknown evils come upon the earth. No genes could have foreseen these evils, but the genuinely altruistic mutations weather them better than the selfish genes. Within-group selection of rational selfishness is self-limiting as overly selfish groups fail to cooperate and lose out in between-group competition to groups with more genuine altruists within them.

Two other related types of secondary gain to the signaler from altruistic costly signaling may contribute to and reinforce the secondary gain to the group as a whole. Altruistic costly signaling can solve the commitment problem in which Robert Frank (1988) and Schelling (1978) are so interested. A costly signaler is someone you might count on cooperating with in the future, and also someone whom you would trust in a one-shot interaction. In addition, altruistic costly signals tend to be more visible, and among humans more talked about, than other kinds. They are broadcast, and so the signaling benefits to the signaler are amplified.[89] Altruistic punishment might probably be a highly visible, dangerous and therefore highly committed, and highly prosocial signal. These benefits to the signaler come *after* the qualities that the original selectors (for example, the sexual selectors) treat as a signal are established.[90]

Remember that costly, therefore honest, prosocial signals are signals whose bearers one can rely on and trust. As Robert Frank (1988) insists, a hard-to-fake signal of trustworthiness yields secondary gain to the signaler. With Caesar it is better to be fat than lean and hungry, because fat men signal their prosociality. Cassius instead signals the fact that he is self-interested, hungry only for his own benefits and not for the gregarious interaction that makes one Falstaffian. It's worth unpacking Caesar's slightly odd simile: Cassius is hungry but not for food, otherwise he would not be lean but fat. His leanness shows his rapacious power and his rational self-control (he may, in fact, be lean because he is rationally avoiding too much food even while hungry), while fat men like Falstaff place gregariousness above sheer accumulation of political control. Hamlet too is fat, despite his desire to signal his own despair by wearing black, and on this reading his fatness stands for his emotional reliability—the reliable predictability of his emotional responses.[91] Claudius capitalizes on this reliability by accurately predicting that Hamlet will accept Laertes's challenge.

My claim is that altruistic punishment will often be the most vivid signal of fitness, hence a signal we are peculiarly attuned to admire.[92] Hamlet is a central example of an altruistic punisher, visible, emotionally driven, cost-bearing. He kills Polonius, which is altruistic punishment gone awry. But altruistic punishment gone awry is one of the costs the punisher must bear, as Hamlet realizes when he tells his mother that

> For this same lord,
> I do repent: but heaven hath pleased it so,
> To punish me with this and this with me,
> That I must be their scourge and minister.
> I will bestow him, and will answer well
> The death I gave him. So, again, good night.
> I must be cruel, only to be kind.
>
> (Shakespeare, *Hamlet* 2001: III.iv.174–180)

We can add characters who range from Benedick, bashfully forced into altruistic punishment in order to court Beatrice, to Timon and Coriolanus, where altruistic punishment becomes recognizable as spite proper,[93] but spite that succeeds in having an element of the admirable within it (as it never seems to in real life, at least when we call it spite). Margaret (in *Richard III*) and Aaron (in *Titus Andronicus*) are perhaps the first of Shakespeare's impressively spiteful figures, though their trajectories are different: Margaret pays the costs much later than she derives the benefits of altruistic punishment. And while Aaron's spite does not quite conform to the biological definition, since it is in fact a convincing signal that he will put his son's interests above his own, and is not genuinely altruistic, it *does* conform to the biological motivations for spite, which in the long run and the right context enhance reproductive success. Aaron provides a good illustration of spite from a human point of view, although from a genetic point of view his genes are directly protecting their replication. Iago, of course, could be regarded as spite personified. His "motiveless malignity" leads to his own childless downfall, a downfall that weighs little with him in his plotting, since he is so nonchalant about plotting an escape from the damage his vengefulness causes.

Malvolio's spiteful exit, on the other hand, does very little punishing (Olivia feels somewhat abashed by it) at very great cost.[94] Nevertheless, all the spitefulness contained in the name and actions of Malvolio aims at courtship. (This is perhaps Iago's aim too.) In *Twelfth Night,* however, Toby and Maria's altruistic punishment of Malvolio works better as a mutual signal of sexual desirability than his does. Their altruism is cooperative; his is not.

Altruistic Punishment versus Spite

This leads to a general insight, not only about Shakespeare but also about the complex interactions among strong reciprocators and altruistic punishers. Among those we wish to punish or see punished are bad altruistic punishers. The worst case and most obvious examples are fake or hypocritical punishers, like the self-serving Mrs. Sparsit in Dickens's *Hard Times* (1966). More interesting are altruistic punishers who are mistaken in their judgments, and therefore mistaken in the self-righteous pleasure that is the incentive to punishment. The distinction between "good" punishers and "bad" ones is the distinction I have been programmatically eliding so far but that I will now emphasize. It is the distinction captured in the difference of emotional coloring between the terms *spite* and *altruistic punishment.* As I'll argue in more detail about *King Lear* in Chapter 4 (see pages 172–181), the psychological experience the mind proposes to itself is the same. The terminology speaks rather to the judgment that witnesses make of the punisher's actions. Often these judgments will depend on which group we're rooting for in between-group competitions. One notorious example is the conflict between the rebel and loyal angels in Milton's *Paradise Lost. Both* Christ and Satan are entitled to being considered altruistic punishers. The terminological choice one makes between spite and altruism moralizes one's reading of those actions.

The apparent or experiential difference between spite and altruistic punishment can be large, but their connection is robust. In *Le Misanthrope* (Molière 1965) Alceste responds bitterly to the loss of his case:

> The palpable wrong that this verdict does me
> I shall not appeal—I refuse utterly:
> The iniquitous practice it makes crystal clear
> Will leave to posterity reason to jeer
> At our age's corruption, whose evil has lost me
> Twice ten-thousand francs (as this verdict may cost me).
> But those francs twenty-thousand now give me the right

> To invoke plague on injustice with all of my might;
> And against human nature which has done me this tort, ill
> Wishes to nourish and hatred immortal. (II. 1541–1550)

He could very probably avoid the loss, but he is eager to pay to punish the entire society he lives in, the entire age. And yet even this spite is prosocial, since he is paying to leave testimony of his age's corruption, and he therefore leagues with posterity, makes himself a cooperator with future society in its condemnation of the present. He does this despite *thinking* that he will hate human nature (or the iniquity that characterizes it) forever. He anticipates allies in the future humans who will agree with him about how bad *all* humans are, which shows (as we'll see) that even outrage has a prosocial, pedagogical and benevolent aspect in relation to what it hates.

In Shakespeare in particular the drama often takes the form of a rivalry between two sets of altruistic punishers—between the Portias and the Shylocks. As I have mentioned, we approve of the ultimate willingness of Benedick to commit to *altruistic punishment* of Claudio. We disapprove of Claudio's horrendously *spiteful* treatment of Hero at the altar. Both he and Leonato are being spiteful—they are also humiliating themselves, and Leonato's terrible "Do not live" is spiteful in a genetic sense: he is giving up his best bet at inclusive fitness. Don John too is spiteful—that is, he is a strong reciprocator, and is therefore the charismatic object of his minions' admiration. His exposure and punishment can therefore resolve the play: his bad spite is punished by the cooperative altruism of the good reciprocators, Beatrice and Benedick, and the "swing" reciprocation of Claudio. Perhaps the purest strong reciprocator in the play is Don Pedro, and his conversion to Hero's cause is the moment of inertia for our sense of strong reciprocation going right.

I cite these examples here to suggest what I'll argue in more detail below, that much of the conflict that we track in narrative is a conflict between strong reciprocators. We have varying attitudes toward strong reciprocators, depending on context, character, and goal, and strong reciprocators have varying attitudes toward one another. Strong reciprocation can lead to conflict as well as cooperation; we appreciate altruism but resent spite. We like vindication but hate vindictiveness.

Shakespeare and Dickens are particularly good at developing fraught narrative moments out of conflicts among strong reciprocators, out of the play of vindication and vindictiveness. But I think that some such play and some such balance is at work in most narratives involving a sense of virtue rewarded. Brian Boyd (2001) points to the gratification of being vindicated in Dr. Seuss's *Horton Hears a Who*.

Let me add a reminder here: the fact that altruism tends to elicit admiration and therefore can yield reproductive success doesn't mean that it isn't altruism after all. Altruistic punishment occurs in a wide variety of contexts, many or most of which don't offer signaling benefits—the signal isn't received. Even when it is received, the costs of altruism might well outweigh the benefits of the signal (Grafen 1990b). What makes altruism a *costly* signal is that the altruism is real; successful hypocrisy would be a better strategy than costly signaling. But *societies* in which genuine altruism is valued as a costly signal do better than societies in which conspicuous consumption or hording or greed are valued, though these too may be costly signals (and we don't deny that greed or consumption are genuine, even if they are also signals).

Another way of putting this, and a significant way at that, is to observe that valuing altruism as a costly signal is itself altruistic. It is costly to choose mates on the basis of their costly signals.[95] This cost has several components: the time wasted looking for a signal sufficiently high to choose the signaler as a mate; the fact that costly signalers may have a net fitness reduction because of the costs they pay for the signal—such a reduction then being passed onto the chooser's offspring; the fact that the longer a chooser waits to select a mate, the more she may have to settle for less than she otherwise might have got; and the ancillary fact that when selection is bidirectional (as in some human courtship), the longer the chooser waits to make or accept a selection the more time pressure constrains her to amplify her own costly signal—the acceptance of the time pressure being (as with Achilles wasting his short life) already a signal in itself.

These subtle and reciprocal calculations are illustrated in, for example, Wharton's *The House of Mirth* (1994), which begins with Lily Bart on the verge of having waited too long. It goes without saying that she must signal her desirability by being beautiful and accomplished ("If I were shabby, no one would have me" [Wharton 1994: 33]). More interesting, though, is her acknowledgment that she is "very expensive" (Wharton 1994: 31). She rejects those like Dilworth who can't afford her to *be* very expensive. It's not that she requires a certain level of amenity to be happy—she doesn't and knows she doesn't. It's that she requires someone who can afford to make such amenity possible, so that her expensiveness becomes a costly signal itself, signaling to her whether a candidate can bear those costs. If he doesn't have to ask what she costs, he can afford her. This is the difference in *King Lear* between Burgundy, who demands a dowry, and France, who sees Cordelia as "most rich, being poor." It should be noted too that it is Dillworth's mother (fulfilling the cliché) who prevents him from marrying her, as she attempts to mitigate the very costs to her own genetic heritage that go with having costly signalers as progeny.

Taking altruism as a fitness indicator does less for individuals who select genuine altruists as mates than responding to a signal consisting in the ostentatious display of wealth might. To be an honest signal the waste must be real, and altruistic waste has fewer secondary gains for the individual signaling—and therefore for that individual's offspring, which are also the offspring her partner selects her for—than has a demonstrated capacity to bully, say. This has led some analysts of altruism, most notably Trivers (1971), who wrote the founding papers on the mechanisms of reciprocity, to argue that reciprocity is a maladaptation in a modern world in which one-shot interactions with nonkin are the norm.[96] Trivers believes that eventually altruism will die out. But there are now persuasive arguments that one-shot interactions were surprisingly common in Paleolithic times, that is to say the period most relevant to the evolution of human nature, especially in periods of stress or calamity (Fehr and Henrich 2003), and some researchers have established plausible models in which, because of additive signaling benefits, unconditional altruism outperforms tit-for-tat reciprocity (Lotem, Fishman, and Stone 2003).[97] In addition, in those contemporary societies that seem likely to be closest to smaller primitive societies, people tend to play the ultimatum game most rationally, motivated least in both offering and receiving by questions of fairness (Camerer 2003: 11), which seems to me to support the coevolution thesis of group selection (Henrich and Boyd 2001; Boyd et al. 2003), by which altruism and altruistic punishment, far from being maladaptations, seem to be required for interaction and cooperation in and among larger and more complex groups.[98]

Interestingly enough, one way this controversy is thematized is in Jane Austen's *Mansfield Park* (2003b). The realistically minded Mary

Crawford declares: "There is not one in a hundred of either sex who is not taken in when they marry. Look where I will, I see that it *is* so; and I feel that it *must* be so, when I consider that it is, of all transactions, the one in which people expect most from others, and are least honest themselves" (Austen 2003b: 36). But Austen's happy marriages occur not through cheating but through honest signaling—that is, Fanny and Edmund come to know each other's worth through and through, as do, for example, Emma and Knightly. The idea that you come to know others and *then* marry them, that happy marriage begins with a reliable estimate of the person you are going to marry, is a confirmation of Grafen's view of honest signaling as opposed to Fisher's view, reiterated by Dawkins in the passage I quoted above, that runaway cheating can occur more or less at random, when mate selection is spoofed by a signal that is cheap and easy to fake.[99]

To summarize this argument briefly: We admire altruists because they can afford to be altruists. We might admire any embodiment of genetic privilege and wealth, but we have survived because we have admired those embodiments of genetic fitness who have signaled their capacity through altruism rather than through selfishness. Therefore an admiration for altruism has been selected for—an admiration that *also* has an altruistic component. But altruism could not sustain an evolutionarily stable system without the contribution of altruistic punishers to punish the free-riders who would flourish in a population of purely benevolent altruists. We therefore admire a certain kind of altruist—the altruistic punisher. But we admire such persons first because altruistic punishment is a costly signal, and only *then* because it is the right *kind* of costly signal to sustain the group cooperation that allowed us to survive to admire it.[100] We admire Superman's fight for justice more than Clark Kent's meek benevolence not only because Superman is more powerful than (we think) Clark Kent is—after all, the bad guys are plenty powerful too—but because his altruism takes the form of altruistic punishment, whereas Kent's takes the form of altruistic patience.

I hasten to add, however, that altruistic punishment isn't the only form of strong reciprocity that we may be brought to admire. We admire altruism in general, and the heroes or protagonists of narratives are generally altruistic. They pay costs greater than the benefits they can reasonably hope to reap from the costs. They embrace risk, or help others, or commit to love, as well as redress wrongs.[101] Still, it is very hard to think of protagonists of any interesting story (any story interesting as a story) whose characters don't include some element of altruistic punishment. Consider, for example, even such life-affirming altruists as Leopold Bloom, Mrs. Dalloway, or Esther Summerson, all of whom have occasion and indeed obligation to punish at one time or another. This confirms the idea that our capacity for narrative developed as a way for us to keep track of cooperators, defectors, punishers, and higher-order free-riders and punishers, and to be particularly attentive to those we think should engage in altruistic punishment.

Nevertheless, our capacity for strong reciprocation may come out as a tendency toward altruistic rewarding as well, and in particular we will tend to wish to see altruists rewarded (just as we wish to see defectors punished); such rewarding is what a happy ending looks like. Malvolio is gone, the scapegoat altruistically punished, and now our favorites can marry, both the altruistic punishers (Maria, Sir Toby, perhaps Viola) and the pure rewarders (Olivia and Orsino). One interesting scene of such altruistic reward, unmixed perhaps with altruistic punishment, occurs in the exchange between Oberon and Titania when they meet by moonlight and confront each other for the first time in Shakespeare's *A Midsummer Night's Dream.* Oberon declares his authority over Titania, but she turns tables to complain about his neglect of her:

Oberon: Ill met by moonlight, proud Titania.
Titania: What, jealous Oberon! Fairies, skip hence:
I have forsworn his bed and company.
Oberon: Tarry, rash wanton: am not I thy lord?
Titania: Then I must be thy lady: but I know
When thou hast stolen away from fairy land,
And in the shape of Corin sat all day,
Playing on pipes of corn and versing love
To amorous Phillida. Why art thou here,
Come from the farthest steppe of India?
But that, forsooth, the bouncing Amazon,
Your buskin'd mistress and your warrior love,
To Theseus must be wedded, and you come
To give their bed joy and prosperity.
Oberon: How canst thou thus for shame, Titania,
Glance at my credit with Hippolyta,

Knowing I know thy love to Theseus?
Didst thou not lead him through the glimmering night
From Perigenia, whom he ravished?
And make him with fair Ægle break his faith,
With Ariadne and Antiopa?

Titania: These are the forgeries of jealousy.

(Shakespeare, *A Midsummer Night's Dream* 2001: II.i.60–81)

The mode of jealousy is interesting here: they are jealous of each other *not* for the selfish interest each manifests in their human complements, but for the *altruistic* aid they give. One would expect that Oberon would resent Theseus, who is to marry his mistress and love, and that Titania would likewise resent Hippolyta, given her own love for Theseus. But what each wants instead is what the audience wants: Hippolyta joyfully and prosperously married to Theseus, Theseus happily married to Hippolyta.[102] I take it that Shakespeare is explicitly thematizing and testing audience interest in a happy ending among characters they more or less fall in love with for the space of the play. This moment helps suggest that our interest in fictions has in it a component of altruism that engages us as strong reciprocators.

Thomas Nashe's observations about the audience responses to the Talbot scenes in Shakespeare's *1 Henry VI* suggest something similar. Nashe writes in defense of the patriotic and moral influence of the theater against those who see it as corrupting youth. A hero like Talbot, who was killed by the French, would have approved of theater, he says: "How it would have ioyed braue *Talbot* (the terror of the French) to thinke that after he had lyne two hundred yeares in his Tombe, hee should triumph againe on the Stage, and haue his bones newe embalmed with the teares of ten thousand spectators at least, (at seuerall times) who in the Tragedian that represents his person, imagine they behold him fresh bleeding" (Nashe 1958: 1:212). The rhetoric of this sentence is psychologically acute. It corresponds to a fantasy we often have, in which we take a kind of pleasure in knowing a future that our hero or protagonist did not know. Here the future is that of his own vindication by time and history. It gives us pleasure to think of him observing with joy an audience weeping over his death (and that of his son) after his great and pleasure-conferring success in subduing the French.

Nashe's terminology is Aristotlean here: Talbot causes *terror* to the

French; the English audience *pities* Talbot. Both pity and terror are varieties of altruistic engagement. Talbot is a hero because he is an altruistic punisher. He terrorizes the French to his own great cost, including that of the heroic death of his son. The audience's response to this—weeping—shows an instinct toward strong reciprocity on their part. They are concerned for him, and for the experience he meets, and not for themselves. That reciprocity causes Talbot joy. It is in the mode of anticipation—we anticipate with some pleasure the *idea* that Talbot could come back from the dead and witness such things—and while we ascribe the joy to Talbot it's really our own joy in the fantasy that Nashe refers to here. And our joy is altruistic: we anticipate the pleasure of displaying for Talbot the story of his heroism and the tears of its audience.

A few further points should be noted here. Even Talbot is not presented as "identifying with" his stage avatar. He is an observer, and not to be confused with the Tragedian that represents his person (Burbage). Like Odysseus hearing his own story told by Demodokos, he *sympathizes* with himself.[103] His emotions are different from those of the Odysseus in the story Demodokos tells: they are the emotions of the nth-level reciprocator, and not of the person thus pitied. Sometimes these emotions *may* be the same, but in pity (as opposed to terror, which may be shared with a character) they generally are not. We feel pity when someone feels pain or oppression or grief, only rarely when they feel pity, and perhaps never when they feel self-pity. Our vicarious experience of other people isn't a reflection or attenuated copy of their experience: it is the emotion we feel in tracking their experience. Often we feel this emotion on behalf of them, but this doesn't mean they feel it themselves. The idea of feeling something *on behalf* of someone[104] suggests strong reciprocation, that is, an emotionally committed attitude toward them or their situation in which we are moved to feel and to do things for their sake and not for our own.[105] Those we are most apt to wish to reward are the protectors of the innocent, the altruistic punishers of those who would harm the innocent—in this case, Talbot's young son. (In the play it's not their kinship but their, as it were, extra-familial relation to each other that is so moving; Shakespeare likes it when kin become friends, like Hal and his father, or Edgar and his.)[106]

Nashe imagines such a reward for Talbot, a reward we all cooperate to give him by weeping his fate. This is the reward of the story itself (as Enobarbus will hope to earn "a place in the story" by remaining faithful to Antony). Just as gossip is a type of altruistic punishment (which is what all tattlers feel), since it destroys reputation, heroic narrative can be itself a type of altruistic reward.[107] Narrative registers and keeps track of cooperators and defectors, and in doing so contributes to and embodies our attitudes toward them. Narrators or authors are strong reciprocators when they talk about others, and also very likely costly signalers, especially perhaps when the narratives are in the first person.[108] In the Talbot quotation, Nashe is not quite a costly signaler, but does belong to the circle of strong reciprocators, collaborating and cooperating in rewarding Talbot. In fact, it is believed that the phrase "the terror of the French," which Nashe quotes from *1 Henry VI* (I.4.24), is his own. He was a collaborator in the play, and if he did in fact write the speech to which he now alludes, the moment he quotes is one in which he has Talbot use the phrase in sorrowful irony since it is a phrase that the French are now using to scorn him. Nashe repays that scorn in spades in his fantasy of Talbot's return and spectatorship of his own story.

Our much mediated attitude toward Talbot is one in which his status as an altruistic punisher plays an important role. Perhaps most gripping narratives require both an innocent whom others can cheat and an altruistic protector and punisher, although these have a numberless variety of relations to each other. Talbot and his son derive perhaps from Odysseus and Telemachus, or Aeneas and Pallas, and are a dim presentiment of Prospero and Miranda: she as innocent suffers with those she sees suffer, and he as punisher makes them suffer even at the considerable cost of her unconditional love and respect, and of his own magic.

We can contrast the relation between Talbot and his son to that between Dom Casmurro and his own son, Ezequiel, in Machado de Assis's extraordinary novel *Dom Casmurro* (1998). Dom Casmurro loves his wife and son deeply, until the sudden death of his best friend. Aroused to jealousy by his wife's grief over his friend, he becomes aware that that friend is living again in his son, Ezequiel, and his love for Ezequiel turns to hatred. The novel is a first person narrative and in the end fills us with the same frustrated rage against its narrator as Shakespeare does against Othello. (Indeed, Dom Casmurro compares himself to Othello, noting that he has the greater motive since his own wife is guilty.) We are filled with rage because Ezequiel is the sweetest and loveliest of children, and loves his father unconditionally, and that love provokes pure antipathy and hatred in the father, who finally poisons his son's coffee. Our revulsion at this is in no way affected by some biological intuition of the just demands of paternity, and I will claim that no fully human response, from any culture, no matter how honor-oriented, could justify Dom Casmurro in this narrative. The indignation we feel is a good index of our attitudes toward behavior we witness among others. We don't evaluate them in terms of their genetic interests (pace Darwinist accounts like David and Nannelle Barash's that see narrative as simply describing genetic strategies and tactics)[109] but of their human ones, just as we are genetically disposed to do.[110]

I suggest the following as the conditions of possibility of the kind of interest in narrative that we have.

1. We are very good at monitoring others, and in fact a large number of other people, and how they interact with one another.
2. We particularly approve of strong reciprocity, and disapprove of defection. We disapprove as well of second-order defection as manifested in second-order free-riding, and we want to see nonpunishers corrected.
3. We particularly approve of altruists and strong reciprocators, including altruistic punishers like the heroes of revenge tragedies, detectives, military leaders (think of Henry V's more bloodthirsty directives in Shakespeare), and superheroes, and borderline figures like Frankenstein's monster, but also costly rejecters of family selfishness, as, for example Dido, Hermia, Romeo and Juliet, Cordelia, Clarissa, Huck Finn, and Stephen Dedalus.
4. Our approval of such characters is itself a kind of strong reciprocation, like the audience's weeping for Talbot, or Oberon and Titania rooting for and somehow assisting the human figures they love. Altruists ourselves, we like altruists—and not out of identification but out of altruism. This it is that fills us with such rage at Dom Casmurro, who is motivated only by the selfish-gene principle.

The argument I make in this book is consistent with the best insights of narratology. In particular, it is consistent with what is the simplest account that does justice to the transaction that occurs between a purveyor of narrative and its audience. This transaction involves several discriminable agents or quasi-agents. At the extremes are the purveyor of narrative and the audience for narrative. Between them are the narrator, the characters that narrative is about, and the narratee. There is much to say about the importance of the narratee in all narrative. Here I will confine myself to claiming that the last level of vicarious experience is our experience of the narratee's experience of the narrative. Thus the narratee is the penultimate term in this hierarchy of monitoring, which extends from target characters in a story to the characters who monitor them, to the narrator who monitors all to the narratee to the audience.

This is a conceptual hierarchy rather than an actual one, since on the whole the monitoring relation between almost any two terms can be easily reversed, even in the course of the narrative. The narrator can assess the narratee's response, a storyteller or an actor can read his or her audience, a narrator can turn out to be an agent that a character monitors, and so on. This conforms to the idea that any cooperator could also be an altruistic punisher, or fail to be one, or at any iteration become an nth-order altruistic punisher, even if he or she has failed to punish on some other iteration. The simplest way of putting it is to say we monitor those who monitor us. We assess their assessments of us, assessments that may include multiple nestings (the Jamesian wilderness of mirrors in which we assess their assessments of our assessments of their assessments, and so on).

One of the merits of this theory is that it allows for a homology between our response to narrative and narrative's own insights into the motives of human behavior. Because we have a disposition to punish those who don't have a disposition to punish cheaters, narrative can both represent some of these interactions and at the same time appeal to our own tendency to interact in such a way. Effective narratives are therefore likely to be accurate representations of human interactions, just because genuine human interactions are what we are so attuned to monitor. Those genuine human actions themselves have to do with how people monitor one another. Therefore our monitoring of how they monitor one another is thematized in the way they monitor one another, and the way they monitor one another models how we monitor them.

This claim justifies the procedure I will try to follow in the remaining chapters. I will assume that audiences will be able to understand and respond only to characters motivated by the kinds of emotions, passions, and desires that we track in the complex negotiations of cooperation, defection, and nth-order punishment in which we are immersed. Narrative has an effect on us in much the same way that characters have an effect on one another. Altruistic punishment is something we approve of, and indeed something we anticipate with pleasure, and I think this is the pleasure of fiction. The altruistic punisher will tend to be a hero, not because we identify with such a punisher but because we like anticipating punishment, and are disposed to require those who detect defectors to punish them, even at their own cost, and we therefore are disposed to wish to reward (approve of, cheer on, root for, love) someone who does detect and does punish.

Any argument of the sort I am making relies on the idea that there is something that we can recognize as human nature. Much other armchair evolutionary psychology tends to see human nature as hard-wired at a significantly more specialized level than I am doing, and tends to be correspondingly reductive. I believe, in contrast to people such as Sugiyama or Joseph Carroll or E. O. Wilson, that narratives are capable of a very large variety of subject, form, theme, and presentation. But our interest in narratives will nevertheless always depend on our emotional recognition of motive and desert among characters.

Successful stories appeal to such recognition, and for this reason they can also be used, as I use them, to illustrate features of evolutionary psychology. We feel as strongly as we do about various characters or situations in stories because of the propensity with which nature has endowed us to respond to such characters and situations, or to such presentations of character and situation. Conversely, any good storyteller will appeal to such propensities and his or her stories can therefore be used to illuminate and analyze them. I attempt such a dialectical account in this book.

I need to say another word about how universal I want to make my language about how narrative works. I assume throughout this book that most of us respond in the same way, despite the fact that many of the models of strong reciprocity I have summarized might seem to resist the notion of a single human nature. Most models of evolutionarily stable systems containing strong reciprocators anticipate a stable mix of strong reciprocators and rational calculators. The latter would defect if they could, but the existence of strong reciprocators makes defecting a rationally bad idea. In the mathematical accounts that model cooperation, defection, and punishment, these practices are generally assigned to three different *types* of individual. In real life it's more likely that most of us respond and behave in all the ways that could be described by each of these terms at different times or in different proportions.[111] Since the aim of the mathematics is to model probabilities, there's no reason to assign every individual to only one of these general categories; we could talk about individuals in terms of their probability of acting in one or another way. In that case, there's a sense in which we could all understand motivations that we need only sometimes feel. We are likely to respond differently at different times, and in fact Lotem, Fishman, and Stone (2003) model tendencies toward reciprocity without assuming that members of a population are stable in their responses over a lifetime. Groups where there is a greater tendency to cooperate and to punish noncooperators probably contain individuals who in any given situation would be more likely to be strong reciprocators. Some will do so more than others, but a varying mix of opportunism and altruism can be thought of as common to almost all descendents of surviving Paleolithic populations, and this is perhaps why we all have some interest in narrative. The variability of people is something that narrative need not reflect, however, because every strong emotional response to a person places that person, for the nonce, in one of the categories to which we respond. We classify strangers as cooperators, defectors, and strong reciprocators, and characters in fiction are strangers to us, and therefore it is natural thus to classify them.[112]

I have sought to establish a way to parallel the interests of audience and agent in the incidents and actions that narrative recounts without using the theory of identification. We are interested because, like them, we monitor cooperation and altruism, where such monitoring is itself cooperative and altruistic (which is exactly why we praise their monitoring). In particular, we look for altruistic punishment. All agents within a social group monitor such things, whether for their own benefit or altruistically, and all of us, more or less, take pleasure in contemplating altruistic punishment and the vindication of genuine cooperators it provides; as social creatures this is the grand elementary principle of pleasure in which we live and move and have our being. We all approve of altruists (even criminals root for the good guys in movies)[113] whether we're genuine altruists or not, and to the extent that we are altruists we follow with interest the actions of others, more particularly the extent to which *they* accurately monitor the cooperation and altruism of those they interact with. We're like them as monitors, and monitors with an innate interest in seeing altruism victorious. But we're not like them because we identify with them. For the audience, the pleasures of fiction are the pleasures that belong to an audience.

2

Signaling

> Advertisements are now so numerous that they are very negligently perused, and it is therefore become necessary to gain attention by magnificence of promises and by eloquence sometimes sublime and sometimes pathetick.
>
> —Dr. Samuel Johnson 1759: 145

In Chapter 1 I tried to rehearse the arguments that altruism in general and altruistic punishment in particular are costly and thus honest signals of fitness, that altruism, and in particular altruistic punishment, therefore confer long-term benefits to individuals and to groups of such individuals who are disposed to altruism, not because altruism aims at such benefits but because it doesn't. (A way of putting this is: cheating yields worse outcomes for someone than not cheating, because altruists punish cheaters, but *only* because altruists punish cheaters. Without their altruism, cheating wouldn't yield worse outcomes.)

I will now consolidate a claim about the relationship among costly signaling, altruistic punishment, and vicarious experience. In Chapter 1 I treated signaling, altruistic punishment, and vicarious experience (in the form of monitoring cheaters and the reactions of other monitors of cheaters) as more or less independent components of an argument for the development of our capacity and desire to monitor the nonactual, and to monitor the monitors of the nonactual, sometimes at many levels of hierarchy. Now I wish to consider how in humans these three components affect one another, and come into a relation with one another in which signaling, altruism and altruistic punishment, and vicarious experience all contain one another as an important part of their respective mechanisms. They are all facets of one another, or all facets of the same thing—a thing that makes fiction, or interest in the nonactual, possible.

In this chapter I'll discuss signaling in order to spell out some aspects of altruistic punishment and vicarious experience, especially the way they are treated or the way they occur within fiction; later chapters will discuss altruistic punishment and vicarious experience more directly.

Two Types of Signaling

The matador kneels down, his back to the bull. He displays this *adorno* because he knows that the bull will not charge. The crowd cheers. The bull is disoriented, weakened, confused, and demoralized. When the matador kneels, he is showing the bull his supreme confidence. As Amotz and Avishag Zahavi (1997) have argued, signals of strength tend to take the form of studiously vulnerable postures. Male dogs and wolves lift their legs to urinate, standing unstably on three legs to claim and mark territory. Human males puff out their chests and lean back on their heals—the opposite of an effective fighting stance.

What is the matador confident of? That the bull will recognize that confidence. If it doesn't, the bull will still be able to kill him. This is a costly signal on the matador's part: the cost is in the high risk he takes of death. His contemptuous confidence signals the bull that his charges and snorting have been vain. It's an honest signal as well, correctly warning the bull away from wasting further energy on charging. Not that the alternative will be much better for the bull, whose exhausted compliance will still lead to its death. But the *signal* itself, the matador's disdain, says nothing about the consequences of compliance, only about the fact that the bull's aggression is unavailing.

A number of issues should be distinguished here. The first one: honest signaling doesn't by itself suggest anything on the order of altruism. Warning a rival away may require considerable energy, as in the vastly energetic roaring or snorting that some animals undertake. This is one of the ways it may be costly.[1] Because it is costly it is honest. To waste energy shows that you can afford to waste it and still fight; if, in fact, you're bluffing about your own prowess as a fighter it takes only one call to lead to potentially deadly consequences. This means that dishonest animals won't reproduce. Among good poker players *true bluffs* (as opposed to strategic bluffing) are very rare.[2] Although the bluffs may not be called much, when they are called the cost to the bluffer is greater than the sum of the much smaller benefits that may have accrued to him in previous rounds of bluffing; in addition, one or two called bluffs might give him a reputation as a bluffer, a reputation that will henceforth deny him even those small benefits.[3] But wasting energy is easier than hurting a rival even when an animal can do so, just as getting an opponent to fold, even when you have a pretty strong hand, is better in the long run than playing a deal out to the end, given all the things that might happen with some significant degree of probability. So an honest warning to a rival isn't altruistic; it's self-interested.

In the case of the matador, that self-interest has to do with getting the bull to be more compliant, so that the matador can kill him more easily. The signal he gives the bull is honest but partial, and in no way altruistic. It is honest but it deceives the bull as well, since he is tricked into acting in ways even worse for him. We should therefore reiterate the principle that while altruism can be a costly signal, costly signals need by no means be altruistic.

The matador's studied disdain for the bull isn't a signal only to the bull, however. It is also a signal to the cheering crowd: a signal of the same thing—his superiority—but not to the same purpose. The crowd cheers the fact that he can turn his back on the bull. The matador is vividly visible to the crowd, and impresses them with his death-defying skill, grace, and courage. He does defy death, and it's real death that he defies since a large number of matadors have been gored and many killed. He does this for the crowd, for the glory—certainly not for the money, that can't possibly compensate for the extreme risks of being a matador. But, as Hume (1978) points out, to do something for glory is *not* self-interested; the desire for glory is essentially counter to self-interest.

As we've seen, it isn't counter to the interests of the matador's genes, however, since the display he broadcasts in fighting the bull is universally reported to contribute to his sexual attractiveness. From a Darwinian perspective, the bullfighter is engaged in a courtship display, demonstrating his fitness through the costs he is able to bear: through his death-defying skill, grace, and courage. He may face a considerable probability of being killed, but he also may well be compensated for this by having many more sexual opportunities—opportunities to reproduce—than his average male compatriot. We can summarize some of the argument that I rehearsed in the first chapter here by noting that populations in which courtship consists of bullfighting might do less well in the long run than populations in which it would consist of firefighting, and that is how genuine altruism might develop and sustain itself. (I mean this heuristically, of course, and not as a comment about actual contemporary societies.)

I want to stress something else here: the distinction between broadcast and focused signals. The matador signals both the bull and the crowd. One of the things that he's broadcasting to the crowd is what he's signaling to the bull. Some signals may be only broadcast, some only focused. The tightrope walker signals the crowd; the boxer, feinting and bobbing and weaving, signals only his opponent (even if the crowd is watching).[4] Different signals have different functions and are variously designed to ensure publicity sometimes, secrecy at others, or they may be indifferent to publicity just so long as the intended addressee receives the signal. The boxer is more or less indifferent to the crowd's appreciation of his signals to his opponent; the bullfighter publicizes his signal to the bull, but also keeps secret *from the bull* the full extent of the vulnerability that he publicizes to the crowd in turning his back.[5] His cape is red—a color that signals conspicuously to the crowd and not to the color-blind bull, which responds to the cape's motion, not its color. Very likely his spangles, which again signal conspicuous and courageous vividness to the crowd, are read by the bull more as a swarm than a single antagonist; the bull goes for the cape that seems solid, and not the spangled clothing that doesn't. Face-saving compromises often have such a structure as well: a private signal allows for cooperation between a stronger and a weaker party, while a somewhat different signal is made public.

Likewise the Zahavis (1997) distinguish between the coloration of lions, more or less camouflaged at all distances, and that of leopards, whose spots tend to camouflage them at greater distances but make them conspicuous close-up. This interplay of invisibility and vividness allows leopards to get close enough to a herd of prey for the signal of their presence to be of mutual benefit. The herd will bolt, and their stotting or energetic signaling when they bolt will indicate to the leopard which animal stots least and is weakest. That animal is therefore sacrificed by the herd itself—to the benefit of both leopard and other animals in the herd. The leopard, then, is concerned to signal but not to broadcast its signal, so that it can signal the prey at the optimal moment for its own hunting, and take down the animal that signals the least likelihood of escaping.

The broadcasting of some signals, however, is a highly striking phenomenon. The Zahavis give a brilliant reinterpretation of the idea of "warning cries" among the babblers they study and among vervet monkeys (1997: 77–80). Babblers and monkeys respond appropriately to different kinds of cries made by others in their group. Monkeys have different calls for the approach of an eagle, a leopard, or a snake, and other monkeys respond correctly to these calls, for example, avoiding the ground when they hear a monkey give the snake call, avoiding exposure to the sky when they hear another monkey make the eagle call, and so on. Babblers have particular calls for flying raptors, for diving raptors, for waiting raptors, and for snakes as well, calls that other babblers respond to appropriately. The Zahavis argue that the calls are *directed at the predators*. They tell the predator: I see you and am ready to take the appropriate action should you try to get me. Both do better when prey signals predator that it knows the predator is there. The predator will go off and look for unprepared prey, and the prey will be spared the costs of flight as well as the small but real risk of being caught and killed if being prepared isn't enough.

Other members of the prey animals respond to these broadcast calls as eavesdroppers. If a monkey calls to a snake something like, "I see you, snake, and am ready to climb out of reach," this call will alert other monkeys to this fact, although that is not the intention of the call. What some researchers have understood (to use speech act terminology) as an illocutionary utterance—*in* making this call I act to warn you that there's a snake in the grass—the Zahavis regard as a perlocutionary effect: *by* informing the snake that I see it, I happen to alert you to this fact as well. Other animals derive from the signals focused on the predator the information that the predator is there; their own response can repeat and propagate the process. One decisive piece of evidence for the Zahavis' views is that signals that have the effect of alerting others to the presence of a predator are not intraspecific: "Many animals react appropriately even to the calls of other species, which are clearly not directed at them" (1997: 80).

My interest in these signals again has to do with the way humans seem to have a special and complex capacity for sending, managing, checking up on, and evaluating different kinds of signals. If you think of altruism (including altruistic punishment) as being such a signal, then the relation of the broadcast and focused aspects of the signal becomes highly suggestive for a theory of vicarious experience, and in particular the experience of narratives.

Let me lay out the general outlines by way of a central scenario. The costly punisher teaches the defector a lesson, and also teaches all observers of the punishment a lesson. In addition, this certifies to observers that he or she has not been a second-order free-rider but has cooperated with others in punishing a defector. The punisher has therefore sent a narrowcast signal to the defector, and a broadcast signal to other members of a group whose members monitor one another. That act of signaling has several components. Defectors and those contemplating defection learn the lesson not to defect lest they incur punishment. Those who perceive the punisher's observation of the defector's defection observe as well that he or she punishes defection. In signaling the punisher is concerned about the reception of the signals and of the message he or she sends. Has the punisher successfully punished the defector? Has the defector learned his or her lesson? And have those monitoring the punishment of the defector concluded that the punisher has successfully communicated to the defector? The altruistic punisher is successful when observers monitor his or her successful communication to defectors, and therefore is intent on observing both the defector's and the monitors' reception of the signal, and the monitors' observation of the defector's reception of the signal. In sending these signals the punisher is concerned about their reception—about whether others will see that he or she has done what was necessary. A component of that is the question of whether the defector has actually received the message, and whether others can see that he or she has received the message.

This is an example of the nested structure of vicarious experience that occurs regularly in real life and in literature. In real life people often complete the action of altruistic punishment by certifying or ratifying it, perhaps most often through narrative, as when a punisher tells a vivid story of how he or she has taken revenge or humiliated or shown up a defector. Such stories may be at once part of the process of punishment (through the devaluation of the defector's reputation) and an exhibition of the altruistic punishment others demand to see displayed (which enhances the punisher's reputation).

We can see here one prime motivation to gossip—gossip is most often a form of punishment, one not counter to Foucault's (1977) sense of distributed nodes of surveillance and control. Axelrod (1997) too suggests that gossip acts as a way to punish defectors even, or especially, when the gossiper is not an interested party. The pleasures of gossip, like the pleasures of revenge, might then be the reward mechanism by which natural selection disposes us to the work of monitoring, narrating, and following narrations about the relations between people with whom we may not otherwise be much concerned. This may be why we're much less comfortable gossiping about kin than about nonkin. From this perspective gossip should be seen as a mode of altruism—have I got a story for you!—even if psychologically it feels like self-preening viciousness.

Proust (1987) gives a quick account of the dialectic of gossiper and subject of gossip in one of the narrator's first analyses of Andrée. He is ambivalent about her tact. She never repeats others' comments about him, or about anyone, and while this is perhaps praiseworthy as tact, as a kind of self-denial, it also means that she dissimulates, and is therefore not trustworthy (2.276–277). She's not trustworthy because she doesn't cooperate in the great social project of making people visible to themselves and to one another. This fact goes to the heart of the hero's central question in the book: What does Andrée know about Albertine, what is she keeping from him, and why? Andrée's dissimulation and persistent refusal to clarify the truth about Albertine's sexual desire and proclivities make it impossible for the hero to know in the end what to tell us about her, make it impossible to resolve the narrative-within-a-narrative that would be the hero's story of Albertine. (The story Proust tells is the narrator's and not Albertine's, and includes the narrator's biologically altruistic bitterness at not being able to tell Albertine's.)

If gossip is a central element in the altruistic imposition and regulation of social norms, then we can see in it a natural motive for bearing tales and for listening to the tales told. We can also discover whence the opprobrium we feel toward a vicious gossip, that is, toward a person who ruins reputation unjustly. This is because the kind of action that occurs in altruistic punishment may be difficult to tell from noncooperation (as with Alceste in *Le Misanthrope*). Part of what makes it altruistic is how easily it can go wrong; the punisher risks being confuted and then seeming vicious and self-serving (or what we tend to call spiteful).[6] The punisher has to have a reason to punish, namely that the individual punished really is a defector. Otherwise it's not punishment but selfishness. Indeed, intentional false witness or vicious and unfounded gossip might be understood as a dishonest signal of the gossiper's altruism. If so, the structure of honest signaling systems will need to impose a cost greater than the advantage of such signals. Get caught in dishonest signaling and you will pay more than such signals can afford you in the long run.

Malevolent tongues will suffer because people will gossip about people who slander. They gossip about vicious gossips, gossip about the viciousness of the stories they tell, and if true gossip is altruistic punishment, then gossip about slander punishes false altruism or ersatz cooperation. This dynamic helps to underscore the point I want to stress here. One of the motives both for narrating and for attending to narratives is the element of altruistic punishment within narratives. Narrative punishes a villain or a defector. This is perhaps one reason that stories of hypocrisy exposed are so interesting. The storyteller is an altruistic punisher, sometimes in the story (how I got them to refund the widow's money) and always, perhaps, by telling the story, since the publication of the story undercuts the villain's or defector's reputation. The reputation that gossip in particular undercuts tends to be a reputation for cooperation (what else do we gossip about except secret betrayal or self-serving actions?). In other kinds of stories we might tell, we might narrate the unmasking of a defector, with the explosion of his or her reputation and the power, status, or authority that derive from it, so that a story describing the villain's comeuppance might simply act to amplify or intensify whatever it is the story is already doing as a story to someone's reputation. The comeuppance would simply confirm the truth the narrative presents. Such confirmation is particularly gratifying because for a species so gregarious and social as we are, a major component of losing an encounter (and thereby suffering in one's reputation) is almost always the fact of being known to have lost. We do not care whether Kasparov loses practice games to Deep Blue when he's at home; we do care when he loses the public contest. The public contest is what it most concerned him to win, what he most sought to win, and so public losses are properly amplified by the very fact that they occur in public. This too confirms the prosocial impulses of narrative.

Because it is the case that from the perspective of the audience, the very publication of the antagonist's loss is an element of that loss (even if the loss is just his or her exposure as an antagonist or defector), the audience participates in altruistic punishment as well. The readership of a newspaper is ipso facto, just by existing as a readership and without any intervention or activity of its own, just by knowing, the punisher of the scandalous behavior that the newspaper exposes.

At least as fundamental as our attitudes toward the subject of gossip is our attitude toward the gossipers, since we are second-order monitors of the altruistic punishment undertaken by those gossipers in their narratives. One reason to attend to narrative is that we are constituted to be interested in whether others are engaging in altruistic punishment—are reporting on bad behavior. Narrative is related to such acts of punishment in two ways, then: through its often gratifying content (when the story is about a defector discomfited) and through the very fact that the gratifying facts are made known. The publication of punishment adds to the punishment, and makes it all the more gratifying. To these reasons to be interested in narrative we should also add the further motive that we will be disposed to pay attention to narrative because we will ourselves want to monitor how the storytellers contribute to the cooperative project by monitoring and punishing defectors. The story tells a story of punishment; the story punishes as story; the storyteller represents him- or herself as an altruistic punisher by telling it.

I emphasize here that punishment of the antagonist is only one motive for narration, the kind most fully found in gossip or the cultivation of outrage. In speech act theory this might be summarized in the idea that just telling the story punishes the antagonist. The story doesn't have to represent him or her as punished, only as exposed, by the story itself. Such exposure doesn't represent punishment; it is punishment.

But such punishment seems peculiarly hollow when it comes to nonactual beings—beings either purely fictional or no longer actual. We now know, let's say, that the Rosenbergs were guilty. I don't think this fact by itself allows for much pleasure in their exposure, since they're dead, but it no doubt allows those who maintained their guilt to jeer at those who didn't. Such jeering is an example of the punishment of nonpunishers (those who, by maintaining their innocence, refused to punish the Rosenbergs). This interaction between triumphalist statements of "I told you so" and disappointed defenders leads to a second principle: we wish to observe the reactions of the guilty to their exposure. We wish not only to expose them but to make them know that they've been exposed. The Rosenbergs were able to maintain their innocence, even beyond the grave, until the Venona decrypts and the Soviet archives were opened, and so they died without ever having the proof of their guilt made gross, palpable, and irrefutable. This fact may disappoint us, and may disappoint their deceived and disappointed defenders especially.

There are two related but independent points to be drawn from our sense that as far as punishment goes exposure may be an important facet in the gratification we feel or propose to ourselves (as I'll suggest in Chapter 4 on vindication and vindictiveness), but is not by itself sufficient to gratify our anger. First, pure exposure isn't enough for narratives of the nonactual to be gratifying, not enough for fiction pure and simple. Very few satisfying narratives take the form of someone getting away with something until some ending by which the narrator simply announces or demonstrates that the person is evil. Most such narratives are failures: the way Shakespeare's *Troilus and Cressida* debunks Achilles might be an example of the best that can be done. It may happen in gossip that exposure of someone's bad behavior to an audience alone is enough, but usually that's because we know the people being gossiped about. It doesn't happen in fiction. We need more than our own knowledge, more than the narrative's exposure of a cheater to us. Any narrative that shows us someone getting away with something heinous is the opposite of gratifying. Our own knowledge of someone's guilt isn't enough. We also need an exposer, an agent who exposes the defector within the narrative, and not only the narrator who exposes the defector from the outside, through the exposition. We need a Holmes or Poirot, not only a Doyle or Christie; a punisher or protagonist *in* the story, and not simply the exposure of an antagonist by the story. This is because what counts as punishment is exposure to the social world in which the defector lives, not exposure to us—the omniscient narrator and omniscient audience—who don't interact with that world. When the dead are exposed, we get the gratification of abusing their defenders, but fictional villains have no real defenders of their actual innocence—how could they, since there is no actual innocence or guilt in question?—and so their exposure to us by itself cannot gratify our desire to punish.

This fact militates against Walton's (1990) view that we imagine ourselves observing what occurs in a fiction. If we were observing, then the fiction's exposure of the culprit to us could be enough to satisfy our desire to punish him. But it isn't. We just don't play a role in narratives of the completely nonactual, even though in cases of actual defection exposure to us might play a role (as it does in gossip). Our knowledge may be enough when we belong to the same society at the same time as the culprit. But our "make-believe" knowledge, the make-believe knowledge of real persons, is not enough to play a role in the lives of make-believe people. I might add that in fictional situations where the culprit is dead, exposure might be just the opposite of what we want, as at the end of Greene's Brighton Rock (1938), where we are revulsed by the horror ("the worse horror of all") that awaits Rose when she finds a means to listen to the vicious phonograph recording her beloved Pinky made for her. (Her desire to experience the nonactual love she believes he has felt for her itself opens her to the horror to come.)

Second, we need to feel that the exposed culprit knows that he or she has been exposed. And even that isn't enough, if the exposure doesn't have the effect of hurting the culprit in a way commensurable with the pleasure we imagine him taking in successfully cheating. Iago's silence, like bomber Timothy McVeigh's, is maddening, while Claudius's death, caught in his own toils, is gratifying. Iago doesn't suffer from his exposure, whereas Claudius does, and it's this that makes us feel the punishment is serious. We relish the fact that Richard III has bad dreams, bidding him despair and die; his death alone would not be satisfying.

We'll return to these questions in Chapter 4. Here I want to make some preliminary observations about the ways broadcast and focused signals work and about their relation to narrative.

Understanding narrative at all requires understanding of signaling (or of semiotic or expressive behavior). We monitor signals and the reliability of signals that others produce. We take note of how others monitor signals, and what signals they produce in turn on receiving signals that we also may receive. One of the intricate pleasures of narrative, from the detective story to Plautine and Shakespearean comedy to Henry James to Patricia Highsmith, consists in keeping track of who knows what. We like to keep track of what other people are keeping track of.

I have suggested that we do this in order to monitor whether others are responding appropriately to what we know they know, even as others measure whether we respond appropriately to our assessments of how others are responding to what we know they know, and so on. I have argued that this talent and propensity is essential to human cooperation. Narrative relies on the psychological incentives to engage in such monitoring of how we respond to what we know about one another.

Overhearing

The simplest definition of drama is overheard speech. In drama, and in other narrative forms, the audience keeps track of how characters are responding to and evaluating one another. Often these responses will consist of evaluations of honest signals. Honest or costly signals are particularly appropriate for narrative tension because they can come into dramatic dissonance with what a character wants another to believe he or she thinks or intends (talk is cheap). We can then observe one character with grounds, good or otherwise, for skepticism about another's stated claims, and can observe the complex reactions such skepticism elicits, and the consequences of these reactions.

We can do this partly because of the fundamental relation between broadcast and focused signaling. We observe the signals characters give to one another just because they are public, but we may also know things that the characters don't know or vice versa. Artful narratives will make sure that focused signals are also broadcast, so that we can interpret or decipher what a particular character is interpreting or deciphering. Narratives (narratives as such, rather than, for example, secret or ciphered messages) by their nature cannot make much use of signals that are not broadcast, since the audience receives the broadcast spectrum alone. Of a fiction, the public can know only what the author or work makes public. Narratives are themselves broadcast.[7]

I will illustrate the following claims:

1. People signal to one another, whether they intend to or not. Sometimes they signal against their own intent, conveying truths that they mean to keep hidden or at least making their own falsehoods detectable.
2. Honest signals are trustworthy ones, and we all tend to reward the trustworthy and punish defectors, even if they are sometimes inadvertently honest.[8] We do this whether we are immediately concerned in a transaction or not, and we do it not only for first-order cooperators and defectors but also for those second- and later-order monitors who should be "strongly reciprocating" the behavior of first-order cooperators and defectors.[9]
3. We are able to engage in such strong reciprocation because most signals are broadcast, or at least are cast broader than their immediately intended recipients. One result is that we can evaluate the honesty of the signal sent to the recipient.
4. Part of the information a signal contains might be the fact *that* it's a signal (see Kreps and Sobel 1994: 866), or intended as one, and often it's that very fact that is being broadcast. The matador broadcasts the fact that he is signaling the bull; the bull does not know that the matador intends to broadcast this fact—if he did, he might be able to work out that the matador is vulnerable.

5. A number of signals select their receivers on the basis of receiver psychology—including the receiver's socialization. Such receivers will themselves signal their acknowledgment of the signal. Not to signal such acknowledgment can then itself be a giveaway about an individual's psychology or socialization. Gift-exchange always has this structure. Gifts officially signify that they place the recipient under no obligation, but any recipient who fails to perceive the obligation and reciprocate appropriately will be anathematized, not only by the giver but by observers of the transaction. Or to take a related example, it's important to know when "No presents" means "Presents expected."
6. The costliness of signals means that there is a continuum between signal and referent. If the appropriate response to receiving an expensive gift is to give an expensive gift in return, the signal itself will have a value that will be real and not arbitrary. This relationship can be many times mediated, but even so signals are synecdoches of the real thing they offer. The costly peacock's tail is not valueless in itself, since it offers the peahen sons, who in turn will be attractive to other peahens. The system, according to R. A. Fisher (1915), may not in itself be a valuable one, but the values and payoffs within the system are brutally real.
7. Failure to receive or acknowledge a signal therefore can indicate a noncooperator. Indeed, as we've seen in the first chapter, cooperation is its own signal. To paraphrase Robert Frank (1988), the best way to seem to be a cooperator is to be one. How do altruists respond to noncooperators? (Remember that altruism here means altruism to victims of defection, not to defectors.) The answer is: they may seek to exclude or kill them or they may attempt to reform them.

This last point is crucial. For humans signaling involves a large component of vicarious experience. We want other people to know certain things. In particular, we want observers to know our bona fides, and we want defectors to know that they've been caught.

The interesting thing to me is the desire to communicate with the defector. There's a certain kind of anger or outrage that is revealingly self-contradictory: the kind where I want the object of my outrage to share my outrage. I want to say, "Just look at how awful you are!" But what's awful about the person I want to say this to is that he is culpably oblivious to a sense that he's being awful. I want to wring a psychological concession of his awfulness from him (which if he made it would no longer justify my assertion or his concession). "So young and so untender?" is one version of this, as is, later in *King Lear,* "I gave you all!" Lear appeals to the heavens as to impartial observers—"You see me here, you gods, a poor old man"—and it is such an appeal that we all tend to make in our spluttering self-justifications of our outrage. But this appeal *also* tends to be made to the very people who wrong us. We want something impossible: we want them to acknowledge the fact that they are too depraved to acknowledge their manifest depravity.

Such satisfactions are hard to come by, as the *frustration* that is so central a self-sustaining cause and component of anger makes clear. In Chapter 4 I will give some account of how literature can compensate for them. But we can see that exposure goes some way toward meeting our conflicting attitudes toward the malefactor: that he is so impervious to shared human decency as not to share with us our outrage at his imperviousness. There is no way to communicate to a truly evil defector a sense of how evil he is. If you could, it would mitigate the evil. Anger or outrage has a self-contradictory fantasy element within it: we fantasize that someone will be struck to the heart by her own heartlessness. We will succeed in communicating to the evil defector by striking her to the heart. What we will succeed in communicating by striking to the heart is that she has no heart. We want both of these at once. If the evil defector has a heart, our anger falters. If the evil defector is indeed heartless, we can't make her feel the force or justice of our anger.

The kernel of my psychological argument is here: Our intuitive sense of wrath and of punishment contains both components. We wish to cause harm to defectors or culprits or villains, and we also wish them to acknowledge of themselves that they are defectors or culprits or villains. In extreme cases we want something self-contradictory: a full-hearted, anguished acknowledgment from defectors that they are so inhuman as to be incapable of full-heartedness or genuine human anguish.

I want to suggest, tentatively, that this psychological dysfunction within anger may play an interesting set of roles in social relations. It will first, and perhaps most important, lend urgency to the corrective function of anger. It is better to reform than to kill a defector—better for the social group as well as for the punisher (and, of course, for the defector). Reform may not satisfy the desire for revenge, but then maybe nothing can.

Second, someone's obliviousness to outrage or failure to become reformed will make his or her defection less ambiguous and more palpable to the social group of which she is a part, if not to herself. The group's discovery of its mistaken judgment of one of its members satisfies our desire for effective and enlightening communication, while the defector's continuing obstinacy or lack of remorse and contrition satisfies our desire to display her manifest criminality. Irreversible errors in punishment will decrease, and the nature of the defector's noncooperation will be more vivid. The fact that the defector doesn't recognize his or her actions as defection will indicate that the person is incorrigible. Third, the expression of anger is a single and efficient signal that will alert observers that the angry person is punishing a defector (and therefore engaging in prosocial behavior) even as it measures the degree of the defector's culpability and possible corrigibility through the response (from contempt to remorse to legitimate excuse to explanation) that it elicits.

In our role as monitors (potential strong reciprocators) we have an abiding interest in the signals people give one another and the success with which they interpret one another's signals. We can follow those signals because they are broadcast as well as focused. One of the reasons that they're broadcast is to signal the quality of the signaler. As signalers ourselves, we have an abiding interest in how our signals are being received. The reception of our focused signals may probe receiver psychology, eliciting information both for the signaler and for the observers at large to whom it is broadcast; the reception of our own broadcast signals may win us status, sex, food, or other advantages.

It turns out therefore that vicarious interest in the experience of others—both those we interact with and those we don't—is central to human sociability and cooperation. We are interested in the point of view of those with whom we interact, even when we feel antipathy for them, and we are interested in the point of view of those we observe interacting, whether we feel antipathy for them or not. Moreover these two kinds of interest are continuous with each other, since we are interested as observers in whether some individuals correctly interpret the signals that we too are receiving, and interested as signalers in whether observers are receiving the signals that we are also focusing on those with whom we're interacting.

One obvious and familiar example of this triaging component of signaling is irony, which always comes at the expense of those who don't recognize it. Hamlet's irony shows up his victims, sometimes without their knowing it, and helps him (and us) discriminate between how dangerous Polonius is and how dangerous Claudius (who understands most of it) is. As to the voice of the storyteller, Stendhal and Austen provide good examples of narrators who select fit audiences, and perhaps partly flatter audiences into thinking that they are fit. George Eliot's ironies have the effect of correcting readers as well, making us realize that we too are the objects of irony that we first believed ourselves to be merely appreciating. Irony and anger can come together in sarcasm, which tries to communicate to its object the ironic perspective that its object won't understand the sarcasm. This is one place where the idea of communicating with the defector can resolve some of the conceptual inconsistencies in anger—as when you roll your eyes *at* someone and not just about them, as though to make them complicit in your contempt for them.

Again, the relationships and interactions we establish and track can be iterated at metalevels, so that I can express anger about unwarranted anger that I observed, and so on, or roll my eyes about someone's anger, or be angry about someone's rolling their eyes. I can do this whether I have a stake in the interaction or not. Everybody has an opinion, and that's an opinionated thing to say. Perhaps the best way to capture the different levels of observation and evaluation that occur is to observe that anything that can be gossiped about—anything that can be fodder for an interesting story or account of how some people responded to one another—belongs to the self-regulating system of interchanged signals that reach and are interpreted by the gossipers and audiences of the gossip, of the stories being told.

Stories are the medium for much of this dynamic system of monitoring, and they are interesting because we are so tuned to monitoring one another, and how we monitor one another, and what stories we tell about one another. One consequence of this set of claims is the fact that illustrations from literature are very helpful indicators of our ability to track how people interpret one another, just as much as "real-life" illustrations would be (they are real-life illustrations), since in a literary performance both works and audiences are required to enter into an even more complex iteration of the iterated social activity being described (see O'Gorman, Wilson, and Miller 2005). In particular we can see something about the interest and accuracy with which humans monitor the signals and responses their neighbors give one another, by looking at the way narrative appeals to our capacity and propensity for such monitoring.

I am making an argument for the usefulness of an idea of verisimilitude. For a purveyor of stories, whether a gossip or an author, to produce a moment of verisimilitude, he or she must have a sufficient sense of receiver psychology to know what we will recognize in the interactions of the characters in the story. We will recognize social behavior. So moments of verisimilitude in signaling are evidence for how signaling works within social behavior, and what it can indicate when it is broadcast.

This is a way of reiterating Brian Boyd's point (2001: 206) about the centrality of folk psychology to social interaction. Our judgments of verisimilitude appeal to our folk psychology. Narrative relies on and therefore provides evidence for how folk psychology works and what it does. Folk psychology is the true medium of social self-regulation, and we could therefore say that folk psychology in general, and narrative in particular, are modes and methods evolved for us to monitor one another's cooperation and prosocial behavior.

Narratives depict signals, and narratives signal, and in doing both they conform to this general truth about signals: that the fact that they are signals can be an important factor in what they signal, especially when those signals are interpreted by onlookers. When the fact that something is a signal is broadcast, the broadcast will engage an irreducibly vicarious component of our experience: we want to know how some people signal or respond to signals from others.

Because *any* signal has a vicarious component—I want to make you understand something, or convey some information to you—the vicarious experience of onlookers is redoubled. We look to see what the signaler signals, to whom, whether that signal is received, how accurately, and whether the signaler perceives what we perceive about the reception of the signal. We observe other observers as well, to see whether they are responding appropriately to the drama we see them see unfolding. We may become emotionally engaged in all of this, not from personal interest or identification, but insofar as we are constituted as cooperators and strong reciprocators.

Literature seems well suited to allow an account of the complexities of signaling and monitoring signals. In what follows I'll assume without frequent reminders that everything a narrative makes us observe is broadcast to an audience by the narrative, so that we may frequently observe moments of signaling within a narrative that are not meant by the characters doing the signaling to be broadcast, or not broadcast to us.

This fact also cuts against the idea of identification, since in response to signaling we will often strongly reciprocate or volunteer affect not in itself felt or signaled by those we do it for, a fact of great interest to Hume (1975) and Smith (2004) and that features in the arguments of Philip Fisher (2002) and Noël Carroll (1990) (this is an issue we'll return to in due course).

How to Do Things with Tears

I want to give a series of examples of honest signaling, partly to illustrate how it works or how we perceive it working, and partly to suggest how central it is to our understanding of narrative interaction. An obvious example of such signaling occurs in the first part of *Henry IV,* when Hal goes to see his father. Henry wants to rebuke himself, but despite himself he weeps:

> not an eye
> But is a-weary of thy common sight,
> Save mine, which hath desired to see thee more;
> Which now doth that I would not have it do,
> Make blind itself with foolish tenderness.
>
> (Shakespeare, *1 Henry IV* 2001: III.ii.89–93)

Henry's tears are an honest signal of his tenderness for Hal, tenderness that he would not seek only to hide but perhaps seek not to feel. As a supreme rationalist he does not wish to have his judgment distorted by too forgiving a view of Hal; he has already said that he has been too mild. His tenderness is moving because it is an honest signal, and we know it to be an honest signal because he tries and fails to check it.

This may strike one as obvious; it should. The point of the example is to show how transparent honest signals are. They are transparent despite their signaler's immediate desire. Honest signaling has a long-term advantage, and what makes honest signals honest is that they are difficult to fake and difficult to hide, even when doing so would yield a short-term advantage. Emotion commits us to doing things that might be against our short-term interest, and *palpable* emotion declares that commitment. To use Robert Frank's formulation from *Passion within Reason* (1988), visible emotion solves commitment problems. You can trust someone who acts on the basis of passion rather than reason, and you can tell that they are acting on the basis of passion when the expression of their passions is counter to their own immediate interests. This is one of the things that make that expression an honest—and so costly—signal. Edgar nearly gives himself away to Gloucester in the hovel because Lear's plight makes him weep: "My tears begin to take his part so much / They mar my counterfeiting" (Shakespeare, *King Lear* 2001: III.vi.56).[10] He does manage not to give himself away, though Gloucester, it transpires, has unconsciously recognized him. Had Gloucester fully recognized him, he would have prevented Edgar's acknowledged "fault" in keeping his identity secret from Gloucester (which Cavell [2002] analyzes so beautifully). The costly signal would or might have yielded a far better outcome.

What Hal gets out of his father's emotion is a sense of the depth of his father's love for him. Let's stipulate as a Shakespearean motif permanent anxiety on the part of children whether their fathers love them. Shakespeare's children tend to act out as a way to test their fathers' love. Hal and Cordelia might be called alike in this, and we could add Desdemona and Hamlet to the list. (Why did Hamlet's father abandon him? Why did he return? Is Hamlet his instrument or his beloved son?) Edmund and Edgar both belong as well, and we might even add Brutus (as Harold Bloom suggests) if he suspects Caesar is his father. The way Shakespeare's children act out is often by withholding the duty that would express their love. Some of this withholding is long term, like Hal's; some short term but catastrophic, like Cordelia's. What Hal and Cordelia (and Edgar) seek to provoke, what in the end they *do* provoke, is an expression of love despite the will of those from whom they elicit it.[11] Henry's weeping is on a continuum with Gloucester's great lines (which he does not know that Edgar hears) about seeing Tom O'Bedlam the night before:

> I' the last night's storm I such a fellow saw;
> Which made me think a man a worm: my son
> Came then into my mind; and yet my mind
> Was then scarce friends with him: I have heard more since.
>
> (Shakespeare, *King Lear* 2001: IV.i.34–37)

These moments are interesting not only because they are irrepressible signals—like the blushing analyzed by Darwin (1998) and to which we will return—but also because they reveal something to the signaler as well about his or her own character or personality. Henry and Gloucester come to know something about themselves by becoming aware of what they are signaling despite themselves (see Bloom 1998 on Shakespeare's characters as self-overhearers). What they come to know is something like the fact that they have it at heart to communicate with those with whom they think they don't want to communicate. To have this at heart is to learn something about your heart.

Tears, as more or less involuntary signals, are difficult to fake. They are also likely to be costly in a purely biological sense since they do blind the eye that weeps; they might therefore signal a capacity or opportunity for genuinely emotional communication. When I weep my weeping wishes to engage your sense of my sorrow.[12] The urgency with which we weep communicates our helplessness, and the cost of the signal is one of the things for which the signal asks pity.[13]

Tears and anger often go together, and I suggest that they can sometimes signal a critical moment: I can afford to weep because I can afford to put myself in a helpless position. Doing so may engage your sympathy, but if it doesn't, woe to you as well as me when I dry my eyes. We feel that this is a last chance for Hal to keep his father's love, and he takes it. Contrariwise, Lear's refusal to weep denotes the impotence of his anger. He cannot afford to make himself any more vulnerable to them. Not weeping is itself a costly signal—"This heart will break into a thousand pieces or ere I'll weep"—but one that he knows will barely register on his daughters. The fact that it won't register is one of its costs, and that cost instead broadcasts to the rest of us (Kent, the Fool, the audience) the extent to which he has been wronged and the grandeur of his impotence.

Ahab's celebrated tear just before the final battle in *Moby Dick* (Melville 1967) represents a critical moment—the last moment when whatever remains of his humanity can win out over senseless revenge. Its power and drama come from the clarity of its meaning. We can see that Ahab is human, that he has fellow feeling for his crew, and that human altruism might compete with the senselessly altruistic punishment of the whale. That punishment *is* senseless: Moby Dick is maddening just because he is opaque to human rage and human passion. This is what the excremental whiteness of the whale means. Ahab can signal to his fellow beings; he cannot successfully signal to the whale.

Shakespeare's Antony is prone to tears, as Enobarbus admits to Agrippa when they see the far more unusual spectacle of Octavius Caesar close to weeping as he parts from Octavia. Sorrow in a rationalist like Caesar should be a warning to Antony not to spurn his sister, and it is a warning he neglects. Antony's tears, which Enobarbus regards as a convenient price to pay for hypocrisy, will nevertheless later make Enobarbus weep as well when Antony weeps not for his own danger but for that to which he has led his followers. They stand for Antony's sympathy—a sympathy that extends even to his enemies. This is the fundamental contrast between his character and Caesar's. Antony wants Caesar to know that he is "angry with him," but for the Roman rationalist this anger is of no interest whatever. Caesar (like Bullingbrook engineering Richard II's murder) seeks to work without honest signaling at all and so rebukes Agrippa for not taking matters into his own hands and murdering his rivals without an explicit order. Antony signals his magnanimity persistently and at great cost ("I have a ship / Laden with gold, take that, divide it, fly / And make your peace with Caesar"). That magnanimity makes his followers weep, loyal to the costly love and costly sorrow he has allowed to sink him but that ensures that he will never betray his men (unlike Caesar who betrays them whenever convenient).[14]

There is no need to multiply obvious examples, since the point is that they *are* more or less obvious, easy to interpret or understand. Tears exhibit humanity, especially when they are unwilling. We know they are unwilling when they are costly. They may humanize a character—precisely because the character is in a situation where being humanized is just what he or she does not want at the moment. This is what makes them honest or reliable: they stand for a quality that it may not be in the best interests of someone to display, and so they show that person as disposed against his or her own best interests, making it possible to get via honesty what could not be won by strategy.

Remember that costly or honest signals do come at a genuine cost. They may show that signalers can afford to pay those costs, but they don't imply that they *want* to (I don't mean experientially, since they may think they want to; I mean if the motive of the signals were pure and rational biological selfishness). If they wanted to, the costs wouldn't be real. They are of value not because signalers want to signal but because they don't want to.

In the paradoxical world of signaling sometimes the cost born is that of *not* signaling when you can. The strong silent types are attractive (at least among Meriam turtle-hunters) *because* they don't boast, as they are entitled to. Not boasting about their difficult feats is part of the signal to which those difficult feats also contribute, and so not signaling can function as a powerful signal. The same is true in rituals of gift exchange, where the gift-giver disparages the costly object embodying his or her extravagant generosity: "'The remnants of my food today, take them; I bring them to you,' said whilst a precious necklace is being handed over" (Bronislaw Malinowski's *Argonauts of the Western Pacific*, quoted in Mauss 1990: 99).

Often such a dynamic will occur as an interplay between the signal intended for a particular receiver and the signal as broadcast. Caesar's clouded aspect is intended, if for anyone, for Octavia and shows her his love. But it also should serve as a warning to Antony not to betray Octavia if he wishes to stay on good terms with Caesar.

According to the Zahavis (1997: 119–121), crying in infant animals is particularly effective *as* an interplay between focused and broadcast signal. The infant's cries attract the attention of predators, *and are meant to do so.* Its parents must hush it in order to protect not only that infant, but others in the group, including themselves. We are disposed to hate the sound of uncontrollable crying because we know that this is a signal not only to us (to whom it says, "Feed me" or "Comfort me") but also to others (to whom it says, "Eat me" or, at least for human infants, "Get my parents to shut me up").[15] Once again, certain animals attend to the signals other animals are transmitting to predators. However, animals for whom the number of predators is large, and therefore the risks of broadcasting a location very high, don't cry for food.

The continuity of the practice of signaling through loud and unpleasant protest, from birds to mammals, from human infants to human adults, shows the extent to which communication is not primarily identificatory. Crying is an example of what Freud calls magico-phenomenalistic thinking. We want something to be the case, and think that by crying loudly enough it will become the case. When an infant cries it has no psychological intention of signaling the lion where it is. It intends to express desperate hunger to its parents. Such despair seems to the infant an irreducible motive for its parents to feed it. In general, we retain an archaic belief that our desperate passions should be enough for other people: this is why we express rage, or weep, or write love letters. As Wittgenstein (1997), also a natural historian, insists, we don't put ourselves in the place of the receiver of our messages, considering how we would react in his or her place (which is to avoid or to placate the desperately passionate). Almost all our passionate utterance is uncalculated. What we don't calculate is how we really appear to those we appeal to, which we would if we identified with them. The appeal itself should be enough.

Urgency seeks to communicate urgency. But what it signals instead is that the signaler feels that things are urgent enough to bear without reflection the costs of being seen to signal without reflection. The psychology of a receiver does not match that of a signaler, but it doesn't need to. What matters is that the signal should be honest and so allow for a cooperative outcome between signaler and receiver.

Signaling is an act, not the production of a message that the signaler looks at from the receiving end. It's always hard to remember how you would feel if you were the receiver rather than the transmitter of an urgent signal, hard to remember that the receiver does not share the perspective of the signaler. Humans can imagine the perspective of the receiver, but only through a process of conscious and artful reflection, counter to the way we are absorbed by our own perspectives in normal communication. Such artful reflection, which purveyors of narratives (especially tricky ones like mysteries and riddles) may undertake, underscores the fact that audiences don't put themselves in the signalers' position, absorbed as we are in their message.

The fact that emotional signaling is uncalculated ensures its honesty. When adults cry they are putting themselves in a position of weakness, not of strength, even when they know with some more calculating part of the mind that this weakness is counter to what they would calculate as their own immediate interests. Talking to Hal, the supremely rational Henry IV thinks he is talking to an enemy ("my nearest and my dearest enemy," he calls him), and the reason he doesn't want to weep is that he doesn't want to put himself in a weak position with respect to Hal. But he does weep, and one of the results of this is to show Hal that he is indeed shaken and wan with care. Hal comes to know the danger his father is in with respect to real enemies, comes to know that his father is not omnipotent, and Hal is therefore more primed to save his life at the end of the play.

In much the same way Satan's tears in Book 1 of *Paradise Lost* (Milton 1975) say more than any words that he uses can. He tries to outscorn his weeping but fails, and that very failure announces his loyalty to his followers. Satan's sublimity, like Ahab's, consists in part in the passions that he cannot control and that tell more about him than those passions he can control.

Contrariwise, what dismays Hamlet is the extent to which tears *can* be faked and used dishonestly. He begins by asserting that his own tears ("the fruitful river of the eye") are simply the actions that a man might play. But when he and Polonius see the player weeping for Hecuba ("tears in's eyes"), Hamlet is appalled by the passion the player can dream up. The world he lives in is one in which he cannot trust the honesty of any signal.

Costly Selection

That these examples are more or less obvious and easy to interpret is one of their merits. Most costly or honest or involuntary signals had better be easy to interpret, as a condition both for the transparency between humans that cooperation requires in general and for the interpretation of any narrative. The second activity builds upon the first and is a subset of it.

Still there are some interesting consequences even in the simple dynamics I have just sketched. The loyalty Hal finally reveals to his father, by word and deed, is somewhat counter to his own interests, as his father himself is rational enough to calculate. While it makes sense for Hal to defend his own inheritance, he could do so without defending his father's life. And his father's own debility would make it all the easier for Hal to become king sooner rather than later. But his father's debility becomes a costly signal for Hal as well: his loyalty (as every schoolchild knows) to his father ultimately ratifies his qualities as a king when his father dies.[16] Loyalty to his father means fulfilling the vow of his soliloquy at the end of the first scene:

> when this loose behavior I throw off
> And pay the debt I never promised,
> By how much better than my word I am,
> By so much shall I falsify men's hopes.
>
> (Shakespeare, *1 Henry IV* 2001: I.i.198–201)

The witty paradox is worth remarking here: paying an unpromised debt falsifies men's hopes. The hopeful men are his rivals as well as his would-be exploiters (the pun in *falsify* looks to Falstaff), so that throwing off his loose behavior will disappoint them. But both the loose behavior and the paying off of a debt he never explicitly assumed are costly signals, and are meant to show us all, in the end, how much he can afford. One of the things he can afford is to save his father's life (another is to save Falstaff's), and it is this fact that endears him to those who had been suspicious before.

This is an example of the interesting literary phenomenon by which a selector's choice of a costly signaler acts as a costly signal itself. His father is now paying the nearly ruinous costs of his own impressive presence; Hal knows those costs are ruinous and could profit by adding to them and ruining his father. But he doesn't, so his alliance with the costly signaler is one of the most impressive costly signals he displays. Likewise, Satan weeps when he sees his ruined but faithful followers, and they listen mute with attention to their sublimely ruined captain. Each party uses the other as a signal to the other that it is paying the costs of faith to a costly signaler. I already broached (in Chapter 1) the interestingly circular phenomenon of signaling one's own ability to bear the costs of preferring a costly signaler; here I want to flesh it out a little.

Most costly signals, the Zahavis say, have some relation to the beautiful (what they call the esthetic; Zahavis 1997: 52, 223–225). The idea is that it costs to sustain beauty. The tautology that only those who can afford to keep up appearances can afford to keep up appearances means that the keeping up of appearances has a useful correlation with fitness. But—in literature at any rate—the sublime is also a part of the esthetic, and I would describe an alternative mode of signaling in which the signs of the world's ravages signal that the bearer of those signs has managed to sustain and survive them. The beautiful and the ravaged are both attractive, and both advertise underlying capacity through the costs their bearers have had to meet (as in Shelley's description of the hero of *Alastor:* "as if that pale and wasted human form / Had been an elemental god").

The apparently straightforward idea that if you pick a beautiful mate you are choosing someone with high underlying qualities is not quite so straightforward as it seems, since there is a cost to maintaining beauty (outweighed though it is by the benefits of doing so). There is therefore a cost in choosing a beautiful mate, as well as in being one, since your descendents will also be likely to be paying the cost of maintaining beauty, again, even if the benefits outweigh the costs.[17] One of the benefits, as we've seen, is that of the structure of mediation: for your genes to prosper, if you have the opportunity to choose you can and should pick a mate whose offspring will be attractive to those who have their own choice. We are disposed to find attractive what we anticipate will be attractive to others, in particular to our offspring's potential mates. It matters that the signal is not going to be received by us alone. We are aware of the broadcast nature of the costly signal, and the fact that it is broadcast enters into our response.

With the sublime, the trade-off of cost and advertising is more obvious, and also slightly more complex. If we're attracted to those with a taste for the sublime—for the dark or dangerous or uncertain—we're attracted to those with expensive tastes. A taste for the sublime is itself sublime, and thus to be attracted to someone with a taste for the sublime is to be attracted to the sublime in that person. Here the costs of such attraction are clear, since we are paying costs parallel to those paid by the attractive figure with that expensive taste. Both sides signal through their costly tastes.

The result allows for a kind of looped but closed circuit. Antony and Cleopatra, Bogart and Bacall, Bonnie and Clyde, Mickey and Mallory Knox (in the movie *Natural Born Killers*) are attractive to each other because each is willing to pay the high costs of loving the other. The cost is high because to love someone who loves Antony turns out to cost more than Cleopatra's other admirers—Dolabella, or Thidias for example—could possibly pay. Only Antony can pay it. And to love someone who loves Cleopatra is a price that only Cleopatra could pay: Fulvia and Octavia cannot. To take a less happy example, Desdemona loves Othello for the dangers he has passed, and he loves her because she pitied them. Each is attracted by the high quality advertised by their object choice itself—by their choice of the other. But this attraction is the opposite of pure reciprocation. Othello doesn't love Desdemona because she loves him; he loves her because she is attracted to danger, because she is the kind of person who loves someone *like* him (as she herself hints to him). She likes what the danger he has passed shows about him, and he likes her for liking someone dangerous. (And we know that danger is real.)

We could make the general point by saying that people may be two things for each other: objects of love and costly signals. In Shakespeare's *Much Ado about Nothing*, Don Pedro, somewhat ruefully makes Beatrice a courtly offer:

Don Pedro: Will you have me, lady?
Beatrice: No, my lord, unless I might have another for working-days: your grace is too costly to wear every day. But, I beseech your grace, pardon me: I was born to speak all mirth and no matter.

(Shakespeare 2001: II.292–296)

There is some truth to her joke (as to all her bittersweet jokes): her own origin and sadness as an orphan make it the case that he is too costly a mate for her own character to sustain. This is a way of saying what all feel immediately, that she could not remain Beatrice *and* marry Don Pedro. Her character (our sense of who she is) could not sustain the cost.

More banally, we're accustomed to talking about trophy spouses or arm candy. But think of it from the point of view of the trophies. They've selected partners who can afford them, who have the requisite resources of wealth or power or status or some other component of fitness. The "trophies" are themselves signals that are broadcast at large of their partners' value or status or fitness. This is what makes them trophies—but their own trophies are the partners who can afford them, who can take them as trophies. Their own status is enhanced by choosing partners who can afford to use them as signals. Everyone recognizes the meaning of the stereotype—recognizes trophy spouses at a glance. Their fitness denotes their partners' fitness and therefore indirectly their own. Through what they signal about their partners, their fitness represents itself.

Even Dorothea's marriage to Causabon in *Middlemarch* (Eliot 1977) conforms to this behavior (behavior that Dorothea herself might easily and blamelessly understand). Something like this dynamic is described in Pausanius's speech in the *Symposium* (Plato 1997) on what the nature of the beloved says about the lover. In Proust, who is perhaps thinking about Pausanius's speech, such a subtle dialectic of signaling is partly behind Odette's relation to Swann, and recapitulated more clearly in the relation between Morel and the Baron de Charlus. Morel, just by being Morel, advertises the fact that Charlus can afford him.

A final turn of the screw comes with the idea that the partner then might use the trophy's youth, beauty, or virility as a costly signal to the youth himself that the partner can afford the trophy. This might be a way of describing Osmond's relation to Isabel: the quality he honestly advertises (and that she misreads) through a kind of disdain even for her is so powerful that he can dispense with Touchett's wealth and with Lord Warburton's status. Indeed, the excesses of love are themselves a costly signal of the capacity to afford them, and those excesses take the form of an urgency whose banner and object are one. I announce to you that a voice in my heart keeps repeating, "You, you, you." You are the signal I offer you of my desperately costly and reliable love for you. (Socrates reproves just this dynamic in the *Phaedrus,* Plato [1961]) The costs of the chosen erotic object are also a signal to the object that one can bear those costs, so that a person becomes at once object of desire, signal of the desirer's advertised sense of self-worth, and addressee of the very signal that he himself is, the signal of the desire for him whose force the signaler wants him to feel.[18]

Note that in such an account of love, there's a circuit through the social or public domain.[19] The qualities of Antony and Cleopatra are widely known. Only Antony can afford to love Cleopatra, and the quality that his love for her advertises, makes Antony attractive to her—and to everyone else. Only Cleopatra can afford to love Antony, and the quality of soul that her love for him exhibits makes her attractive to him—and to everyone else. The costly signal is broadcast, and we can be magnetized by them without being willing or able to pay the costs of dedicating ourselves to their catastrophic desires and actions. This relation to the costs they advertise and demand from their partners is what allows us to be attracted to them, fascinated by them, without identifying or wishing to identify with their partners. We love them from afar; we can't afford anything else, and so our psychological interest in them can be purely spectatorial. Narrative works because, in general, we can monitor dazzling and attractive signaling that we would not be able to pay the costs of responding to.[20]

This fits in with the idea of monitoring and strong reciprocation. We want to see them happy, but this does not reduce to the idea that they are our own object choices. They are dear to us though too dear for us. We are interested in their happiness, but not interested or able to stake our own happiness to make them happy. In such scenarios, each lover thinks of the other just as we do. What makes them grand is that they are willing to pay the price of loving each other, the price of taking each other as objects of love. We love and admire them for that. They love each other for that. We love them for loving each other.[21]

As I say, this structure is more obvious (though perhaps not more powerful) in its sublime manifestations. Satan weeps for the millions of spirits who continue to follow him, even when his attempts to address them are broken by weeping. And the rebel angels all preferred the costs of Satan's tortured sublimity to the far less costly choice of the loyalists' grace and beauty. Whatever you feel about the rebel angels, what makes them grand is the choice they made to associate with the other grand rebel angels.

Such an account of signaling makes it possible to see how the structure of narrative could work. The signals are broadcast but the responses are much narrower. Those who respond are those who can afford to. Since response is therefore a broadcast signal as well, one that it would cost to respond to, we can follow the signals, feel the psychological impulses they elicit, see figures as simultaneously subjects and objects to themselves and each other, without ourselves moving beyond spectatorship, beyond our own attitudes toward what we want to see others reward or punish.[22]

Our evolved response to costly signals is to want to see them elicit whatever it is we want to see reward them—love, punishment, comfort, food, revenge, and so on. We want to see strong reciprocity: our own tendencies toward strong reciprocity begin as monitoring others for it. We don't have to be the responders but we do want to see the proper response, and this is our basic emotional attitude toward narrative.

Narrative can, of course, thematize this structure as well by showing characters with the same emotional attitude toward the other characters in a story that we have toward either those other characters, or toward all of them. Signals can be broadcast within a work and that broadcast may be remarked upon. When Antony blushes, Cleopatra takes it that "that blush is Caesar's homager, or else Fulvia's." Why does he blush? Blushing is an honest signal of how one feels, of self-consciousness. It is honest because we would suppress blushing if we could, and it shows that we want to suppress it and can't. The obvious ways that blushing can commit people in erotic situations are fodder for literature, and I won't rehearse many here. I just want to note that the reason they are fodder for *literature* or representational narrative is that blushing is broadcast too. We can see, from outside the interaction, someone blush at what someone else has done or said. Cleopatra can announce to her audience, and to the play's audience, that Antony is blushing.

In Shakespeare's *King Lear,* Gloucester (according to his own report) has often acknowledged Edmund by blushing. He not only blushes when he acknowledges Edmund; it is his blushing *that* acknowledges Edmund, and that shows us all that Gloucester is more committed to his son than Edmund is to his father. Signals that we would suppress if we could are honest because we can't suppress them. Raphael's most candid, winning moment is his blush at Adam's question about the angels' love. Lily Bart's blush tells Selden (and us) how she feels about him (Wharton 1994: 77). Proust has St. Loup blush, in an exemplary moment of guilessness, when the hero's grandmother makes him a gift of some Proudhon letters she has bought:

> He was overcome by a joy of which he was no more the master than of a physical state which is produced without any intervention of the will, he became red as a child who's just been punished, and my grandmother was much more touched to see all the efforts that he had made (unsuccessfully) to contain the joy which so shook him, than by all the expressions of gratitude that he might have been able to proffer her. (1987: 2.221)

His blush is a guarantor of the emotion he feels, of his genuine gratitude—gratitude and not only joy, because the fact that he seeks to contain that joy shows that he's not entirely given over to a selfish pleasure but full of good feeling for the bringer of that pleasure. And she is in turn full of good feeling for him, because of his joy and because of the blush with which he seeks to suppress that joy.

In a more subtle interaction, M. de Norpois tells the hero that he will put in a good word for him with the Swanns. The hero suppresses a move to kiss Norpois out of gratitude. Much later we learn that Norpois has seen and understood the gesture, and having concluded that the hero is insane, did not use his good offices on the hero's behalf. The lesson for the hero is one in dissimulation: signaling *is* costly, and those costs can lead to losses in social situations. The frankness they signal, however, is a net gain in the long run, but only if the frankness is real and likely to be costly in individual situations.

In Proust the abstract drama of signaling informs the whole book, as its secretive hero becomes more and more successful at dissembling his emotions and sustaining his opacity, even as he discovers more and more the damage that such opacity causes both himself and others. He does this by becoming a better and better reader of the sometimes very subtle signals of others, as when he understands Andrée's suppressed ambivalence about his relation with Albertine.

To clarify matters here a little bit I want to summarize the way that Robert Frank (1988) puts George Ainslie's extremely important work on hyperbolic discounting (Ainslie 1975) to subtle and powerful use.[23] Hyperbolic discounting is a term describing the way we value immediate benefits much more highly than far-off ones. Most of us will take a gift of $100 today instead of $105 a week from now, but most of us will prefer waiting fifty-two weeks for a gift of $105 to taking $100 in fifty-one weeks. Our sense of future value drops off in a logarithmic perspective. A possible explanation for this somewhat irrational psychology is that our short-term and long-term interests (I want just one more cigarette; I really have to quit) often diverge. We discount the future because we are individually geared to deal with the dangers of the present. We can't reproduce if we don't survive.

But our genes can't survive if we don't reproduce. We therefore have two potentially competing interests, or perhaps instincts or agencies, at work within us: to survive and to reproduce. They are in a Prisoner's Dilemma situation with each other: if we act so as to survive under the pressure of the moment, we will ruin our long-term reputation and our prospects for reproductive success. But if we look only to reproductive success, we are more likely to die in the present moment; the percentages that are good for our genes may not be good for us. (My genes would prefer me to take a 99 percent chance of immediate death to my getting a 100 percent effective vasectomy; I don't agree with them.) It is possible that we attain a cooperative equilibrium through a kind of agreement where I indulge in short-term discounting of future gains in the "present present" exchange for giving up such indulgence in a future present. My short-term self cooperates with my long-term (or genetic) interests by giving up a later short-term gain; my long-term self (or genetic make-up) allows me a short-term gain now, in exchange for that longer-term cooperation. (There is intriguing evidence that economists, who valorize rational choice very highly, expect and display high rates of defection in Prisoner's Dilemma and ultimatum games; see Frank 1988 and also Camerer 2003: Chapter 2).

Our genes try to get more than this, however, and Frank argues that economically irrational emotional incentives might have evolved in humans to counter our own disposition toward the short-term. Emotions elicit possibly self-destructive short-term behavior that are in our genes' interest but not in our own. They make us likely to cooperate when our own hyperbolically discounting perspective would tend to put short-term gain over long-term rationality. This seems to be an argument against the actual biological altruism of the emotions, but in the complex dialectical system of evolutionary development, it turns out (as we've seen) that my genes must solve the Prisoner's Dilemma (with respect, not to the closed system comprising myself and my genes, but to other people) by constituting me to be a genuine altruist, rather than rational maximizer of *either* my own *or* my genes' interests. Their interests are best served, under the circumstances, in a cooperative and therefore self-monitoring society when they dispose me toward genuine altruism, both with respect to other people and with respect to themselves. Our psychology gives us certain emotional responses to an action we might take or suffer that are counter to our immediate rational interests. We might reject an ultimatum, to our own harm. *Being known* through hard-to-fake or costly or honest signaling to have the emotional propensity to act against our own rational interests helps those who receive our signals to solve the problem of whether they can trust us. Blushing, weeping, flushing with rage, going livid with shock: all these are reliable signals, not only of how we feel in a certain situation but of the fact that we generally emit reliable signals. It pays to be fathomable. People tend to trust those who blush easily.[24]

This analogy might help: Let's say that being alive is like being in a friendly poker game. Every new game is a new generation with the analagous result that every player in a later game is a "descendent" of the player he or she is in the current game. From my restricted *perspective of a player in a single game*, I want to win the poker game that I am playing, and leave richer than I arrived—and I am more likely to win if I am skillful, a large component of which skill involves keeping a perfect poker face. So I might do really well playing if I can hide my intentions and fool the other players. But I am much less likely to get invited back than if I give away a lot of my hands through "tells," difficult to suppress and difficult to fake expressions of glee or disappointment or anxiety. It's in the interest of future generations of poker games, and of the players of those future games whom we have called the descendents of the players of the current game, to be more honest than would be warranted by maximizing the profit in this game. But it's still important to win, at least some of the time, since if a player loses everything he or she won't be able to return to the table. By analogy, losing everything is an individual organism's failing to survive; losing a place at a future game is a failure of reproductive success. Long-term success (doing okay in many games) requires a balance between these two competing imperatives, whereby the profit-seeking instincts of the current player are subverted by honest and prosocial tells. Indeed, the pleasure of poker is in the way it pits honest against manipulative and against play-acted signaling, and while professional players can do very well against amateurs by not giving their hands away (by defecting, in the terminology of Prisoner's Dilemma games), to do so they need a large supply of amateurs who like playing poker. Such a large supply exists only if there are a lot of (pro)social players of genuinely friendly games, a small percentage of whom think to try their hands at the more dangerous but much sparser tables—the niches of the professionals.

Blushing and tears are canonical examples of difficult-to-fake signals, as is smiling: it turns out to be very hard to simulate a convincing smile (Seabright 2004; Robert Frank 1988). These signals are mediated through the emotions that we are framed to experience, often against our own rational interests, in the situations that elicit them. This fact suggests as well that signals can inform one about one's own commitments.[25] Think of all those stories that contain some variation of the sentence, "She felt herself coloring" (a sentence to be found more or less verbatim in Henry James, many times in Bret Harte, Patricia Wentworth, Anne Perry, countless drugstore novels, and countless novels for young adults). Henry IV realizes that he loves Hal more than he wished to admit to himself (he has, after all, said that he wished to discover that Hal and Hotspur had been exchanged in the cradle). Hamlet may realize that he loves his father less than he has hoped when he fails to produce the gall that should make oppression bitter. (Hamlet also seeks to elicit an unwilling and revelatory signal from Claudius at the Mousetrap—"if he but blench / I know my course.") Lily Bart, in *The House of Mirth* (Wharton 1994), learns what she herself thinks of Selden only when she blushes.

A Midsummer Night's Dream thematizes vicarious experience (under the tutelage of Oberon and Titania) and does so partly by thematizing the extent to which the circuitous and self-utilizing pathways taken by signaling can affect receivers. Let's stipulate that the play is all about our concern with seeing others come to grateful and graceful resolutions, with seeing Jack and Jill reconciled. Oberon and Titania are audiences to the drama between Theseus and Hippolyta, and their own tendencies toward altruistic aid to their protégées lead to tension between them (as we've seen in the previous chapter). At the end of the play comes an incident that juxtaposes the idea of signaling, both broadcast and focused, with the idea of theater, that is of storytelling in which we overhear and are able to interpret the exchange of broadcast signals, and produce some of the responses these signals mean to elicit. In Act V Hippolyta, who is just beginning to trust Theseus, asks him not to humiliate the mechanicals. Theseus replies with a reason she can trust him:

> *Hippolyta:* I love not to see wretchedness o'er charged
> And duty in his service perishing.
> *Theseus:* Why, gentle sweet, you shall see no such thing.
> *Hippolyta:* He says they can do nothing in this kind.
> *Theseus:* The kinder we, to give them thanks for nothing.
> Our sport shall be to take what they mistake:
> And what poor duty cannot do, noble respect
> Takes it in might, not merit.
> Where I have come, great clerks have purposed
> To greet me with premeditated welcomes;
> Where I have seen them shiver and look pale,
> Make periods in the midst of sentences,
> Throttle their practised accent in their fears
> And in conclusion dumbly have broke off,
> Not paying me a welcome. Trust me, sweet,
> Out of this silence yet I pick'd a welcome;
> And in the modesty of fearful duty
> I read as much as from the rattling tongue
> Of saucy and audacious eloquence.
> Love, therefore, and tongue-tied simplicity
> In least speak most, to my capacity.
> (Shakespeare 2001: V.i:89–109)

When eloquence is nonchalant about Theseus's presence, it is "saucy and audacious." It signals nonchalance, makes it obvious that the clerks are not sufficiently submissive to Theseus. The very fact the eloquence is difficult to fake makes it a reliable signal of unanxious artfulness, just as the stammering and tongue-tied anxiety that is its complement reliably signals the clerks' actual sense of Theseus's dominance. Their failure to suppress their anxiety is a costly signal of how they really feel. (Remember, it's good for their favor in the eyes of Theseus, just *because* they think it isn't: they *want* to fake appropriate eloquence and can't, which is what saves them; this is what I mean by saying a signal is self-utilizing.)

Moreover, Theseus's approval of the clerks is a signal as well. They have practiced their speeches to be worthy of him. They want to be worthy of him not only so that their praise is grateful to his ear but also so that he will admire them for the eloquence of that praise. His interest in their eloquence would also not be merely the pleasure of the speech, but the sense that those who placed themselves at his disposal were themselves accomplished and effective people. By preferring the tongue-tied simplicity of the fearful clerks (and of the mechanicals) Theseus himself is paying a price. He allies himself with those who are less effective, less powerful, less "saucy and audacious" than he is entitled to demand.

He can afford to do so and signals that fact by picking a welcome out of their silence. This signal, and the underlying personal qualities of patience, kindness, generosity, insight, and reciprocation it indicates, help win Hippolyta's love. She comes to love him because of the way he responds to the signals of others—because of the kind of strong reciprocation he undertakes. In the drier terminology I've been using, we could say that Hippolyta monitors the way Theseus monitors and responds to the signals that the stammerers send him, and which she is able to monitor as well (by report or by performance). Hippolyta's second-order observation of Theseus, of what he observes and how he responds to what he observes, makes her approve of Theseus. Indeed, Theseus displays his response to her as an element of his courtship: "Trust me, sweet."

Again, it's important to remember that he's displaying an honest response. That he *knows* that stammering is a welcome is itself his signal. Contrast this, for example, with Chad Newsome's caddishly clueless assurance to Strether about his feeling for Madame de Vionnet in Henry James's *The Ambassadors:*

> "Of course I really never forget, night or day, what I owe her. I owe her everything. I give you my word of honour," he frankly rang out, "that I'm not a bit tired of her." Strether at this only gave him a stare: the way youth could express itself was again and again a wonder. He meant no harm, though he might after all be capable of much; yet he spoke of being "tired" of her almost as he might have spoken of being tired of roast mutton for dinner. "She has never for a moment yet bored me—never been wanting, as the cleverest women sometimes are, in tact. She has

> never talked about her tact—as even they too sometimes talk; but she has always had it. She has never had it more"—he handsomely made the point—"than just lately." And he scrupulously went further. "She has never been anything I could call a burden." (1909: 312–313)

Chad's failure to recognize her value (even as he crudely signals his vanity in thinking himself a subtle reader of signals like the tactful display of tact) is also a failure to recognize the cost a real recognition of her value would require of him (fidelity to her). The fact that he "meant no harm" shows that he is unaware at this point of the things he might mean, and might signal himself as meaning, which are in fact beyond him. Because he does not demonstrate that he knows her value, his response reliably tags him as a cad. This means that she bears all the burden of her own value: she does it for him as well as for herself, and his very impenetrability becomes a feature of what is impressive (and heartrending) about her. Strether's admiration for her is increased by the fact that she pays the costs of loving Chad; her capacity to pay those costs while still never wanting in tact distinguishes her from any other woman Chad might take up with (even her own daughter), any woman whom Chad would probably have the capacity to describe correctly. After all, who else would take up with him? Only someone radiantly transcendent, like Madame de Vionnet, or else someone at Chad's low level. Chad recognizes her tact, but radically undervalues it, seeing it as a kind of exquisite refinement of manners, when for her it is rather the extreme high cost that she pays for him. It is this moment that confirms how out of love Strether falls with Chad, even as he falls more deeply in love with Madame de Vionnet.

Theseus, however, understands silence. Silence is tact, and tact is a kind of silence, and he meets the clerks' silence tactfully, and in showing that he knows what their silence means he wins Hippolyta. The response that she observes *we* observe as well—since that's what it means when something is public knowledge, available to observation. And we therefore too are won to Theseus, at length. Won to him, we wish to see him happy, tend to strong (that is, altruistic) reciprocation ourselves, and in doing so we want what Titania wants, which is to see Hippolyta understand him and love him.

That she does love him makes it clear as well that she does understand Theseus, much as he has understood the clerks, and we complete the circle by feeling benevolently toward her as well: first for pitying the mechanicals, and then for understanding Theseus's desire to see them perform as tact and not as sadism. So we are won to her and in wishing to see her happy, we want what Oberon wants.

Note too that in choosing the mechanicals (and in picking a welcome out of the clerks' silence) Theseus is also paying a price. He could have seen a better play, as he could have the loyalty of more effective clerks. But by being willing to pay the price, by taking pleasure in paying it, he demonstrates what he can afford to give up, essentially what he is willing to give up for real friendship, loyalty, and love. And this is what the audience approves in him. (To some extent we will reciprocate, not only by approving but by willingly sitting through the tale of Pyramus and Thisbe as well, and by applauding Oberon and Titania when they bless Theseus and Hippolyta, and by applauding the shadows and forgiving them any offense, after Puck's last speech.)

Silence

> The man who threatens the world is always ridiculous; for the world can easily go on without him, and in a short time will cease to miss him. I have heard of an idiot who used to revenge his vexations by lying all night upon the bridge.
>
> —Dr. Samuel Johnson, *Life of Pope*

In *Much Ado about Nothing,* Don Pedro responds to Beatrice's apology for the joke about his being too costly a mate to her by saying, "Your silence most offends me, and to be merry best becomes you" (Shakespeare 2001: II.i.297–298). Silence offends him because it would indicate her fear of him as someone who would take offense at her jokes. To be the object of her mirth is a price he's glad to pay: he can show her that he likes it, and that he understands it. Although Theseus reads silence as welcome, and Pedro as the opposite, the dynamic is the same: each advertises his capacity to understand the surprising response he elicits. The advertisement is honest because Theseus and Pedro have to understand it to be able to show they understand it. To know there is something to understand is already to understand it.

King Lear mistakes Cordelia's silence; unlike Theseus he fails to give Cordelia thanks for nothing or to see that love in saying least speaks most. I want to look at the first scene of the play as a case of signaling gone awry that we can monitor (just as Kent and France can) because we are so good at seeing how human signaling works, and what its relation to honesty and to cost is, and the complicated apportioning of those costs.

Sometimes the cost of a signal is precisely the fact that it *can* go awry. Signals of disdain, for example, don't cost energy but opportunity. By putting yourself in harm's way you signal your ability to tolerate the risk of the harm, a signal worth paying if it lowers the risk of being attacked sufficiently. That is, it's worth paying if the combination of lowered risk and remaining capacity to repel or escape harm is greater than the higher risk of being attacked if instead you attempt to prepare for harm, even if offset by the greater ability such preparation yields to repel or escape harm in an actual attack. The fitter you are, the more you can rely on residual capacities even at a greater risk of harm; the fact that you are putting yourself at risk signals these greater capacities honestly, since once in a while you'll have to use them, when the signal doesn't deter attack. That might be an example of signaling going awry, and the fact that this may happen means that any signal through risk is costly. (Game theorists call the possibility that things can go wrong a *tremble,* and refer to the way things can go awry when they ought not to as *trembling hand perfection,* a phrase I cite here because it's so evocative.)

Cordelia's refusal to give Lear what he wants is a costly signal. Its cost is obvious: she loses the most opulent third of the kingdom; she loses her father's love; she loses her authority and place in the world; she loses her dowry; she loses her opportunity to marry Burgundy. The cost of the signal also makes it an honest one, and she honestly and fully signals to Kent and to France her intrinsic worth. To the extent that she absorbs these costs for the purposes of sexual selection, she gets the superior mate: the generous France and not the greedy Burgundy. France's superiority is partly signaled by the dialectic that I've looked at above: the fact that he chooses the undowered Cordelia is itself a costly signal of his high quality. He can afford her, by which I mean that he is her appropriate match, because he can understand that she is most rich being poor, most choice forsaken, most loved despised. In effect she picks the man who signals his worth by picking her despite the costs of doing so.

This is a very gratifying scenario, common to folktales. Variations of it may be seen, for example, in Chaucer's "Wife of Bath's Tale" (1989), as well as in the theme of the three caskets in *The Merchant of Venice*. Portia signals her worth by choosing the more or less worthless Bassanio (worthless both monetarily and morally); he in turn and without contradicting his initial worthlessness comes to signal his worth by choosing the costly lead and not the obvious and apparently cost-free wealth of the gold or silver.

Cordelia's signal is not meant for Kent or France, though, but for Lear (and also for her sisters—more obviously in the "Mirror for Magistrates" [Baldwin 1960], one of the play's immediate sources). She is signaling Lear her trust that he will understand her silence as Theseus or as Paulina would. Paulina's great line at the end of *The Winter's Tale,* "I like your silence," makes her a later and more profound avatar of Theseus, and while Cordelia's silence is not the same as that which Paulina praises in Leontes, it nevertheless risks the costs of having nothing glib to say. (Leontes's silence turns out to be a sublimely tactful courtship display as well.)

Cordelia by her silence means to trust Lear. She trusts him to love her. This is to say that she trusts him to understand (as France and Kent do understand) that silence is a signal of trust. Whatever the psychological content of her refusal to flatter him, whatever her sense of embarrassment or rage or self-reliance or independence, those emotions imply her unconscious faith that he will understand why she might evince them. She can think of him only still as a father if she can rely on him to understand her as a child.[26] Goneril and Regan can produce false and cheap signals because they no longer regard him as a father. It costs them very little emotionally to flatter and so they do. They have no emotional commitment to honesty or to being understood by Lear. But Cordelia does have such a commitment, and therefore will not flatter him; she needs to be able to afford the costs of being silent because she needs him to show that he understands her. Only he can indemnify her for those costs, and so her silence is a costly signal soliciting him to pay those costs.

That is to say that she needs him to signal his love by accepting the costs she imposes by her silence. Everyone agrees that what makes us take Lear as a man more sinned against than sinning is the kindness that he shows the Fool (Empson [1951] points out, for example, that Lear almost always addresses him not as "Fool" but as "Boy" or "Lad"). The Fool is for Lear a very costly signal of his own depth of character. Goneril's and even Kent's contempt for the Fool contrasts with Lear's kindness to him. The part of the Fool was almost certainly doubled with that of Cordelia (that is, the parts were written so that one boy actor would play them, and the play's *lines* were written so that the audience would take note that one actor was playing them both). The Fool embodies a signal of Lear's residual tolerance, love, and commitment to truth, even under the pressures of horrendous misjudgment, violent mistreatment, madness, and despair, just as Cordelia's silence or refusal to flatter should have allowed Lear to signal tolerance, understanding, love, and truth. Cordelia uses her silence as a signal of her trust in Lear; Lear too should be using her silence as a signal of his love for her. Understanding it, paying the costs of the public embarrassment of her refusal to say she loves him: this would announce his own power and love better than any grand gift he gives.

Lear's own misapprehension of her silence should be for him (as it is for us) the very sign of the great cost of her silence, maintained as it is in the face of his demands and his irrationality. The cost to her of his irrationality and incomprehension is her signal to him of her great love for him. This is what she wants him to comprehend: that she is silent even in the face of his incomprehension. And while this is a paradoxical and seemingly impossible double-bind, it is, in fact, what the play finally shows happening—but only over a long passage of time. (We'll return to the temporality of vindication in Chapter 4.)

Lear doesn't understand, however, or at least not quickly enough. For him the costly signal of his love is the gift of land that he makes. He gives up his kingship, handicaps himself to the extent of reducing his train to a hundred knights, and thinks to show his love that way alone. Whether or not Cordelia understands Lear's gesture as ratified by its costs, the scene depicts the way signaling can go awry. The way it goes awry is in a kind of misprision of a signal's effect. Cordelia thinks Lear knows that children will insist on their own freedom. She thinks, with Freud (in his great essay "Family Romances" [1909]), that the desire for freedom is itself a tribute to the parent from whom the child seeks to free itself: it shows the child's undamaged desire to impress the parent by being on an equal footing with him. If, like Lear, the parent doesn't have the wisdom to understand this (and who does?), the tribute will be felt instead as wounding and painful. It *is* wounding and painful—that is its cost and what makes accepting it as a tribute a reliable sign of love to the person who offers it.

Shakespeare makes this dynamic clear by having Kent repeat it. Like Cordelia he defies Lear on Lear's behalf, and with some expectation that this defiance will be correctly interpreted as loyalty and love. This moment of communication also goes wrong, but remember that it is one of the costs of costly signaling that such signals *can* go wrong. The risk of misinterpretation has to be real for the fact that someone is willing to run those risks to be significant. Regan and Goneril don't risk misinterpretation, and this is why their flattery is empty.

We as humans are constituted by our own natures to be able to follow this drama of interpretation and misinterpretation, and to want to follow it—to be able to see the extent to which people are using one another as signals of their own relations to one another. We understand—and perhaps because we are all absorbed in monitoring we can understand better than any agent in the drama can—that a person can treat another as simultaneously the recipient of a signal and as the signal she wants him to receive. Lear's rage is the cost Cordelia shows him she is willing to pay as a signal of her trust and love in him. His rage is her signal to him.

We in the audience understand and monitor the signaling and reciprocating relations of those we observe from outside. We do so because we have a propensity toward altruism or strong reciprocation or sympathy, empathy, and absorbed curiosity without identification or personal interest. We monitor others' propensity toward altruism and sympathy. We understand motive through honest signaling. We understand how others signal their motives to one another, and understand the extent to which they are reading one another reliably. These are conditions of possibility for human cooperation, and also conditions of possibility for narrative.

People's interest in one another is always vicarious, and in signaling they don't overcome the fact of separation and vicarious relation; they reinforce it. They do so because they make of the recipients of signals the signals themselves. The other isn't a pure recipient, another myself, but also the very thing that I am exhibiting to him. "See, sons, what things you are," says the dying Henry IV. To use another as a signal to that other I must simultaneously acknowledge her subjectivity and also think of her as an object, as the signal I wish to impress upon her subjectivity. This complex or composite relation to her experience is one where my attitude toward her is that it is *like* something to be her, and being like her is what it is like to be her.[27] My relation to her is therefore irreducibly vicarious: I want her to see herself as I see her, not from my point of view but from her own, so that she can vicariously understand my point of view. Our relation to each other is irreducibly vicarious.

Kent's impulse (and Cordelia's too, I have argued) to make Lear's absurd incomprehension the incontrovertible evidence that will make him comprehend his folly allows us to see the relation between the structure of trust and that of the frustrating anger and impulse to revenge that I noted above. The impulse is self-contradictory, in the moment at least, but powerful all the same. Just *look* at what you're like. Do you *hear* what you're saying? "See, sons, what things you are" (2001: *Henry IV*, Part 2, IV.v.67) expects them to see their own obtuseness. This is what anger always (somewhat incoherently) expects.

That emotion signals an intention to hurt or punish its object. Its incoherence and irrationality make it reliable. I want you to cooperate, and therefore I may act in such a way as to injure or kill you, losing any chance of cooperation, a way that may at least alienate you from me, so that you won't be inclined to cooperate. These risks are the costs of the signal of anger. The costs are real. In anger I assert an honest sense of your incorrigibility—with the purpose of correcting you. But if my anger actually convinces you that I think you're incorrigible, it might seem that it's pointless for you to cooperate or correct the behavior.

Anger commits me emotionally to an ultimatum that it might not be rational for me to make: change or leave. Rationally I should take what I can get from you. But if I refuse, out of spite, then you had better assuage my anger if you want what I can give you. Again the costs are the risk that we'll reject the ultimatums we're giving each other when we're angry at each other, and both come out behind. But the fact that I might act spitefully toward you, if I am honestly feeling spiteful, even though such spite does me harm as well, might induce you to the concessions I want even if I think I no longer want them. A hallmark of anger is a willingness to pay the costs of anger. If I am willing to pay the costs, you can't use my self-interest against me. (—"Honey, the neighbors."—"Who the hell cares!")

This is what Agamemnon discovers in angering Achilles in *The Iliad* (Homer 1990). Achilles has chosen a short but glorious life. His sulking in his tents is as spiteful a refusal of cooperation as can be imagined, since he is wasting the short time he has by not achieving the glory that is his compensation for a short life. Agamemnon cannot expect this attitude from Achilles, and he does not. But so great is Achilles' anger that Agamemnon eventually has to appeal not to his self-interest but to his pride—the very pride that is so counter to his self-interest and, in fact, to its own interests. Pride makes him refrain from the actions that would yield glory. Pride loses him Patrocolus. But eventually pride succeeds in wresting from Agamemnon the concessions Achilles requires. Achilles' tragedy in the *Iliad* is the cost of those concessions.

The point I wish to stress here is the extent to which even the most overbearing and self-involved emotions are also signals. They are directed at those they scorn. Such emotions as scorn are in some ways the model of all irrational emotion: they are directed at swaying those to whom they announce their contemptuous indifference. They turn out to be a kind of dramatic sulking, but without the conscious intention in what we call sulking. Emotions about others therefore have a strong vicarious component toward those others: emotion is about what others think and not only about what I think. They are about how another will respond to me just when I *don't* identify with that other, and when he or she *doesn't* identify with me. Emotion isn't about having the same subjective point of view but about communicating a radically different point of view: the extent of the emotion calibrates the difference.[28]

Volunteered Affect

Because emotional interactions are broadcast signals, spectators can also follow the drama of emotional interaction. And because they are emotions, our experience is vicarious. This doesn't mean that our vicarious experience isn't emotional: our volunteered passion—Philip Fisher's term (2002)—or tendency toward strong reciprocation shows that it is.[29] Fisher is interested in the way our own affect or emotional response may differ from the emotions we're responding to. We may know more or different things than the object of our emotion. We may see a man asleep in a field about to be trod on by a horse, and although he is asleep sympathy will make us run to his aid, despite the fact that we are not sympathizing with how he *does* feel. We *anticipate* and supply the fear that he lacks due to his ignorance of the danger he's in (Hume 1978: 2.2.9; P. Fisher 2002: 142–146; see also N. Carroll 1990: 100, in his argument against the idea that the audience experiences "character identification").

Fisher's brilliant argument for the relation of anticipatory fear to the experience of narrative emphasizes what for Hume is somewhat deemphasized, which is the distinction between spectator and fearful incident. We experience alone (or perhaps through the "register" of another powerless spectator) what the objects of our passion are not experiencing. Our own passion is intensified because we do not derive it from the emotions of a character, but supply it on behalf of that character.

Fisher comes close to making Hume sound like Edmund Burke here (although he contrasts Hume favorably to Burke's view of the sublime).

I want to emphasize a different aspect of volunteered affect, however. For the early Hume, of the *Treatise,* sympathy is egoistic; it has a strong element of identification within it, through the double action of impressions and ideas. All sympathy refers itself to one's own self, to how I would feel in the state of the person with whom I sympathize. Insofar as others resemble me I sympathize with them, and insofar as they resemble me I take pleasure in their sympathy with me.

This second point is important: love of fame (Hume 1978: 2.1.11) derives from the pleasure that thinking about myself and loving myself gives me. Fame for what I approve of in myself allows me access to what Freud will call narcissistic pleasure in myself. Hume here explicitly argues against the idea that the "desire for fame" is an "original instinct." We discriminate both about whose approval we want and what we want approval for. We want the approval of those who remind us of ourselves, being related to us in various ways, and we want approval for what we like ourselves for. We think sympathetically of the love and admiration of those who love and admire us because this sympathy is a new route our self-love can take to the perpetual destination of self. Sympathy is directed toward relations, aspects, reflections, reminders of myself. Hume's theory here more or less anticipates Freud's idea of primary narcissism as the reservoir of the libido with which we invest all those that we love. Since libido is self-cathecting, it is our own libido, derived from the pools of our narcissism, which we in fact love in them. (Such a theory makes his theory of identification possible.)

As some of my earlier citations suggest, Hume implicitly recants much of this argument in the later *Enquiries* (1975). In the *Enquiry Concerning Human Understanding* and the *Enquiry Concerning the Principles of Morals* Hume sees love of fame, ambition, and so on as original propensities of the mind, not derivable from strict self-love: "It is requisite that there be an original propensity of some kind, in order to be a basis to self-love, by giving a relish to the objects of its pursuit; and none more fit for this purpose than benevolence or humanity" (1975: 281). Love of fame is in the second *Enquiry* part of the more general and original motive that consists in our wanting to maintain a certain reputation, both to others and to ourselves. We are irreducibly and constitutively social, our pleasures are social, and our morality derives not from sympathy, whereby others remind me of me, but from benevolence, whereby I care about others and therefore about myself as perceived by others. We look with favor on those who are generous, and we seek a similar reputation ourselves because it is for us a primary and irreducible good to be looked upon with favor by others. Hume notes that our leisure and recreation is primarily social, and that no matter how pleasure-loving our motives, the pleasures we love come from what evolutionary biologists will call altruistic tendencies within us. This is because it pays to be generous; generosity elicits the generous sentiments of others. That may be a good reason to be generous, of course, but only because a more general and conceptually prior generosity already rewards generosity.[30]

This later view of volunteered affect runs somewhat counter to Philip Fisher's (2002) sense of the intensely isolated subject of the vehement passions, of fear especially. Hume's account of benevolence is similar to that of the strong reciprocity I've been rehearsing. We approve of cooperation and therefore approve of those who approve of cooperation and further it; this approval of ours furthers cooperation itself and excites the approval of other cooperators. Because we are benevolent we approve those who are benevolent, not for being like us but for cooperating with us in benevolence. We praise someone for "being a man of honour and humanity," for being a person from whom "Every one, who has any intercourse with him, is sure of *fair* and *kind* treatment" (Hume 1975: 269) and this is due to the general fact that "the sentiments, which arise from humanity, are not only the same in all human creatures, and produce the same approbation or censure; but they also comprehend all human creatures; nor is there any one whose conduct or character is not, by their means, an object to every one of censure or approbation." (Hume 1975: 273; see also Frankfurt 2004: 71).

What we censure or approve in others is just their human benevolence, their capacity for being pleasing or useful, not especially to ourselves but to others, to themselves, and to those with whom they interact. Central to our censure or approbation is the extent to which they too are motivated by the benevolence that stimulates us to take both them and those they interact with as its objects. (Censure here is what will come to be called altruistic punishment, and joins with approval as a mode of strong reciprocity.) Whatever a skeptic thinks of the possibility of actually meeting an honorable man, even he or she must concede that "the preceding delineation or definition of Personal Merit [which we all more or less agree on in general terms] must still retain its evidence and authority: it must still be allowed that every quality of the mind, which is *useful* or *agreeable* to the *person himself* or to *others,* communicates a pleasure to the spectator, engages his esteem, and is admitted under the honourable denomination of virtue or merit" (Hume 1975: 277). Indeed, Hume says we regard even the economic virtues such as "industry, discretion, frugality, secrecy, order, perseverance, forethought, judgement," and so on as merits, and we do so because of our benevolent sense of how they will promote the interests of their possessors.

The kind of volunteered affect that Hume considers in the *Treatise* (1978), for example the fear that I feel on behalf of someone asleep as a horse bears down on him, makes for one kind of narrative, the sort Philip Fisher describes as involving suspense or dramatic irony or some sense of relevant knowledge that we have but that the figure for whom we feel concern does not. (Recall the end of Graham Greene's *Brighton Rock* [1938] when Rose goes off to listen to the recording that the vicious Pinkie has made just before meeting his violent end; she promises herself consolation but we know that he has exploded in invective and hatred for her when he was left alone in the carnival recording booth.) But Hume's later commitment to benevolence rather than sympathy (as he had defined it in the *Treatise*) allows us to volunteer good or ill feeling on behalf of those to whom our relationship it a disinterested one.

Hume argues that our moral judgments, derived from this benevolence toward humanity in general, are themselves general (1975: 274) and not perspectival. I don't approve or disapprove on the basis of how action affects *me;* I judge the action independently of my connection to it, and approve or disapprove of what people do in a generalized sense. My individual presence as an observer has nothing to do with my judgment. Just as when I visualize an object I don't imagine myself *seeing* the object, I only imagine the object itself (Moran 1994), when I approve or disapprove of someone's actions through my conformity to the principle of benevolence toward mankind, I am not concerned in what I approve or disapprove. I am attentive to what others are doing, all absorbed in the general attitude of attention, in what Nagel (1974) calls "the view from nowhere."[31]

This doesn't mean that such feelings are in any way attenuated. Adam Smith (whom Robert Frank [1988] cites often) describes a mode of volunteered feeling in his account of sympathy in the *Theory of Moral Sentiments* (2004), when he describes what he is calling sympathy, and what

I would like to relate to Hume's theory of benevolence rather than what Hume calls sympathy. Smith writes:

> Sympathy . . . does not arise from the view of the passion, as from the situation that excites it. We sometimes feel for another, a passion of which he himself seems to be altogether incapable; because, when we put ourselves in his case, that passion arises in our breast from the imagination, though it does not in his from the reality. We blush for the impudence and rudeness of another, though he himself appears to have no sense of the impropriety of his own behavior; because we cannot help feeling with what confusion we ourselves should be covered, had we behaved in so absurd a manner. (Smith 2004: 6)[32]

Sympathy on this account would be fundamentally vicarious. True, Smith puts this in terms of what we ourselves should feel, but that feeling would be imaginary, not real. And indeed, it turns out what we would feel might be confusion, not shame, since we wouldn't act so impudently and rudely if we did have a sense of impropriety to regulate our behavior. The way we feel about others is sometimes a way we could not feel about ourselves. The affects we volunteer are often affects that those we volunteer them for could not possibly feel themselves, any more than they could know that their beliefs were wrong: "The compassion of the spectator must arise altogether from the consideration of what he himself would feel if he was reduced to the same unhappy situation, and, what perhaps is impossible, was at the same time able to regard it with his present reason and judgment" (Smith 2004: 6). It's not only witnessing impudence that causes us to volunteer affect; witnessing distress will too, especially unconscious distress, as will sensing that someone is incapable of being conscious of the way he or she has been wronged or of the doleful situation he or she is in. Smith provides this subtle illustration:

> What are the pangs of a mother, when she hears the moanings of her infant that during the agony of disease cannot express what it feels? In her idea of what it suffers, she joins, to its real helplessness, her own consciousness of that helplessness, and her own terrors for the unknown consequences of its disorder; and out of all these, forms, for her own sorrow, the most complete image of misery and distress. The infant, however, feels only the uneasiness of the present instant, which can never be great. With regard to the future, it is perfectly secure, and in its thoughtlessness and want of foresight, possesses an antidote

> against fear and anxiety, the great tormentors of the human breast, from which reason and philosophy will, in vain, attempt to defend it, when it grows up to a man. (2004: 6–7)

This exemplary narrative has its own expertise. It illustrates how the mother volunteers anxiety on behalf of her child. But it also causes us to volunteer sympathy for the mother, first because we are like her, whereas she is not like her child. The child doesn't know to be afraid, but we do, and we pity her grief for the child because we too pity the child. We also pity her, feel for her something like the way she feels—although not the way she feels for herself. Still we feel for her what she is not explicitly aware of feeling for herself: her helplessness and terror and misery and distress. But there is more going on here: we have some relief in knowing that the child is not as miserable as she imagines him to be, and more in knowing what the future holds, since the end of the story is that "it grows up to a man." The child's future is secure; Smith's narrative says so. It is secure thanks to *her,* thanks to the care she feels and the care she takes. And yet despite this security, we know too that when the child grows up he will be tormented by fear and anxiety, and so we know that the mother's pains will yield an outcome she is not thinking of: the child's torments when he does become an adult, torments not quite like hers, since we assume that his fear and anxiety will be for himself, not for another, as hers are. We volunteer affect for the mother whose child will one day have sixty or more winters on his head, even as she grieves for a fate we are told at the end of the paragraph will not occur: his death in infancy.[33]

I wish to connect these examples of volunteered affect or strong reciprocity to signaling. The theory of signaling that we've been examining is one in which conveying a message in no way requires the (naive) idea that the sender will offer or allow the receiver the sender's perspective. Or to put it another way, our response to signals may be affectively saturated without our emotions in any way implying our identification with the signaler or our putting ourselves in his or her place. Our general stance toward sociability normally makes human interaction immediately transparent to us, without our having to consider what *we* would mean if we said or did what we observe. Signals are normally immediately transparent to us as well, whether we are their intended receivers or not. When we aren't, their effect on their intended receivers is normally immediately transparent to us. Our response to these transparent negotiations is one generally of disinterested, altruistic approbation of generosity, candidness, or other tendencies toward cooperation, and censure of their opposites. Candidness (honesty) is a characteristic of signals that we approve of, which means that signals have a tendency toward something like altruism or at least those signalers who are not calculating and dishonest will be preferentially selected for by altruistic bystanders. And as we've seen, true altruism is itself a highly charged possible outcome of this tendency to honesty or cooperation in signals, since true altruism signals fitness. We approve of signals of altruism, and such approval is like the altruism that it approves in others. Unselfish feelings about others in part take the form of unselfish feelings toward their unselfish feelings.

Again I want to stress, as I did by quoting Schelling on his Altruists' Dilemma (Chapter 1, note 49) that the dynamic of unselfish approval of others' unselfishness can become self-sustaining, and so irreducible to purely selfish motives. For selfish reasons we become genuinely and henceforth irreducibly unselfish. Evolutionary psychology can tell us how we got to the point of genuine cooperation, which is to say how we transcended its regulative constraints. But once we do get to that point, while evolutionary psychology may still constrain the limits of altruism it no longer determines its forms and functions, and its explanatory ability is much weaker than is imagined by the literal-minded evolutionary literary theorists I demur from in various footnotes.

We unselfishly approve of others' unselfishness, but not because we put ourselves in the position of others like us. We don't so much as consider our resemblance to them when our feelings are affected by theirs. We pity the mourning, for how they feel, when they are grieving for the dead, but our feeling is not the same as theirs, even if we weep with them. As for mourning the dead, Smith (2004) regards sympathy for the dead as the clinching case of a purely vicarious version of sympathy, since the mourner volunteers a feeling that it is impossible one should have were one in the situation of the (dead) person to whom the mourner volunteers it.[34]

Vicarious feeling for others is therefore both a propensity for responding emotionally to the signals of others and (in humans) itself a primary example of such a signal. The mother's vicarious feeling for the distress her infant signals is itself a signal to us (of a different kind of distress, a vicarious distress). Altruists signal their altruism, and react with approval to the altruistic signals of others, while signaling disapproval of the defection of others, including that of second-order free-riders who don't act altruistically in their relation to the altruism or defection of others. They do so both when they are parties to an interaction and when they are disinterested spectators. As spectators, they are moved to side with the more altruistic parties of an interaction—as in the three-party versions of the ultimatum game that I noted earlier.

We observe the extent of altruistic signals, and because the way a signal is received is itself a signal, we observe what the receiver makes of a signal. I mean this both idiomatically and more literally. What Lear makes of Cordelia's signal is a mistake, but what he should make of it is his *own* signal of his love for her and his willingness to advertise that love by taking so little by way of flattery from her. Her costly signal could also be his, as it in fact is France's. We approve of what France makes of her signal—he makes it his own—and disapprove of what Lear does.

Our own monitoring of costly signals and our response to the response of others constitute our own costly and altruistic (rationally disinterested) absorption in the interactions of others. And, of course, we signal as well with the stories we love, a mode of signaling that can range from the simple desire to repeat them to the social capital of our own conspicuous cultural attainments. Knowing a story and, still more, telling a story signals our own the capacities for altruistic interest, affect, and punishment, capacities that the story will represent its characters manifesting in order to appeal to the audience's interest in monitoring these things. Stories recount signals and stories are signals. In the next chapter I want to say a little bit about the relation of the storyteller to these issues.

3

Storytellers and Their Relation to Stories

> Another thing to strive for: reading your history should move the melancholy to laughter, increase the joy of the cheerful, not irritate the simple, fill the clever with admiration for its invention, not give the serious reason to scorn it, and allow the prudent to praise it. In short, keep your eye on the goal of demolishing the ill-founded apparatus of these chivalric books despised by many and praised by some many more, and if you accomplish this, you will have accomplished no small thing.
>
> —Cervantes 2003: 8

I have offered an explanation for interest in narrative that acknowledges parallels between the experience of characters and the experience of an audience, while avoiding talk of identification. In place of identification I've considered the mechanism of "strong reciprocity," that is to say, biologically conditioned altruistic impulses toward other human (and humanlike) beings whom we are in a position to monitor. (The word *reciprocity* in the phrase is something of a misnomer, since so-called strong reciprocity does not expect a return for altruistic behavior: this is what makes the behavior genuinely altruistic.) I want to look at a few examples of how fiction gratifies us, in order to defend the claims I have made and that I summarize briefly here:

1. Both positive and negative feelings toward fictional or other nonactual characters are examples of strong reciprocity, of a biologically or economically disinterested impulse to advance or impede them, to defend, praise, or blame them, to feel relevantly gratified or disappointed by their successes and failures.

2. Such feelings will tend to be elicited not only through our general propensity to care for other humans—the general benevolence toward humanity that Hume attests—but more particularly through our observation of their merit. Merit will finally be found to consist in an altruistic or prosocial disposition. By now it should be clear that what I mean by an altruistic disposition will include not only overt self-sacrifice, like Dorothea Brook's or Clarissa Harlowe's, but also a candid desire to promote human society and sociability, or the society of others who are also prosocial—a disposition like Clarissa Dalloway's. Our positive feelings are toward those who are prosocial, those whose positive feelings are likewise toward those who are prosocial, and who respond negatively to defectors. Our negative feelings are toward defectors and toward second-order free-riders: those who *don't* punish defectors. A disposition to cooperate means a disposition to help the innocent, to punish the guilty, and to reward those who help the innocent and punish the guilty, including those who are guilty of not having a disposition to cooperate. Innocence is itself a prosocial quality since it consists essentially in candor and trustworthiness and the prosocial attitude of trust. Innocence is the simplest form of a prosocial disposition, but it is not the only one. Altruists need not be innocent, but they are on the side of the innocent. To be innocent means to be a cooperator.
3. In narratives we will therefore be disposed to want to see the cooperators triumph over the obstacles set up by defectors of various sorts. Our own propensity toward strong reciprocity will make us root for characters with a propensity for strong reciprocity, not because we judge them as *like us*, or identify with them, but because a disposition to reward cooperators and to punish defectors is itself a central aspect of cooperation.
4. Strong reciprocity will often take the form of altruistic punishment. This may be the predominant form of altruism in narrative, and in general the good guys will at some point engage in some form of altruistic punishment. The heroes and heroines of narrative are those who pay the costs of defending the innocent and who punish defectors. Sometimes they will be aware of the sacrifices they make to do so, sometimes not, but we will be aware of the sacrifices they make. In Jane Austen, such characters as Catherine Moreland or Fanny Price don't mean to cooperate in rebuking their

> more manipulative rivals or false suitors or exploiters, but their essential, uncalculated generosity of soul does cooperate in the events that end in such rebuke of those actuated by mean calculation. But, in any case, one of the important characters for whom we're rooting will often be an altruistic punisher—someone who bears the costs of punishing defectors. Those costs may be subtle or highly mediated, as for example, refusal to allow a defector to compensate for what he has done (as when Silverbridge refuses the compensation offered by the cowardly Major Tifto in Trollope's Palliser novel *The Duke's Children* [1973b]), as though the punisher is not, in fact, exacting punishment. But not to exact retribution can itself be a costly punishment.

Because it is costly, and because bearing those costs is heroic, altruistic punishment is a common characteristic of heroes (as Freud also saw in *Totem and Taboo* [1913]), and we could group in this category not only Achilles, Aeneas, Hamlet, Satan and the Son in *Paradise Lost,* Philip Marlowe, Batman, and the Bride in *Kill Bill* (Uma Thurman), but also Emma Woodhouse, Daniel Deronda, Silverbridge, Milly Theale, and Lambert Strether. Some of these characters pay with their lives, some not; all certainly take on the burden of their own altruistic behavior. Achilles' preference for glory to long life is an exemplary instance of such behavior, since the satisfactions of glory are social satisfactions paid for at the highest personal cost.

This is no paradox. Remember Hume: the pursuit of glory (as of revenge) neglects all considerations of interest. It can be successful only because we admire its pursuer's disdainful indifference to interest, the way he places considerations of what others think of him above material advantage. It is not mean. Achilles cares more about his reputation when dead than about anything such reputation could garner him. His reputation is an end and not a means.

The costs of prosociality signal that anyone willing to bear those costs is prosocial, and we prosocial observers are therefore well disposed to the bearers of those costs. We therefore identify the protagonist in a narrative as the figure whose undertakings are essentially altruistic, recognizing a wider range of behavior as altruistic than we might if we defined it merely as genial well-wishing.

Human sociality, or the cooperative or altruistic disposition of most humans, combines these features: we monitor others, tallying the history of their cooperative behavior; we monitor how others respond or fail to respond to what *they* discover about the history of the cooperative behavior of their fellows; we are moved to punish defectors, even if they do not harm us, and to reward altruists, even if their altruism doesn't benefit us; and we are moved to approve of others who do punish defectors and reward altruism. Much human emotional life consists in and commits us to these responses to the behavior of others, and to their emotions, which impel and guarantee that behavior. Here we have in place all the features needed to explain an interest in narrative.

The explanation I have given may find some confirmation in the kinds of emotions that are particularly to be found within narrative experience. In what follows I want to explore some of these. The affect or emotion we volunteer will sometimes be like the emotions experienced by the characters we feel for (as in terror) and sometimes not (as in pity). Both cases, are equally examples of strong or disinterested reciprocity. We respond emotionally to costly signals, and our emotions can be, but need not be, the same as the signalers'. The emotional concomitant of our strong reciprocity is the emotional experience that we have in experiencing narrative; furthermore, our emotions signal that we are strong reciprocators, and probably the mutual sense of everyone else's emotional response to tragedy helps cement their sense of one another as a community. How could it be otherwise?

Authors and Narrators

Among the strong reciprocators to narrative events are the narrators of those events. We've seen that gossip is a likely mode of altruistic punishment: the scandal monger punishes scandalous behavior, and while the gossiper might get egoistic satisfaction out of it, he or she does so only because he or she is altruistically disposed to get satisfaction out of knowing and telling scandal. Muckraking is work, and even if tattling is fun, it's fun because of an altruistic attitude in the gossiper, who takes pleasure in discovering and imparting information to others. Gossiping, destructive though we often take it to be, is like road rage in bespeaking an essentially prosocial attitude. It gives pleasure to other altruistic punishers—that is, those who like gossip—even as it disciplines those who have violated whatever norms the gossipers are punishing.

Gossipers are narrators, and there's an obvious sense in which narrators or authors get to display punishment and reward to an audience that is altruistically committed to appreciating and applauding and so rewarding such punishment and reward. Readers of Dickens and of Richardson famously demanded prosocial behavior on the part of those authors. Dickens is blamed for killing Little Nell (though forgiven perhaps because he has created her), and forced by readers to make Pip and Estelle live happily ever after. Church bells ring when Pamela marries. And, of course, Dr. Johnson excoriates Shakespeare for the death of Cordelia.

Indeed, Richardson thematizes this issue in *Clarissa* (1932), a work replete, in preface, footnote, and text, with moments in which Richardson tirelessly considers the general question of the relation of vicarious—sometimes called "exemplary"—experience to danger and desire. As usual, it's Belford who brings the issue up within the fiction. In one of his attempts to intercede for Clarissa he makes the following suggestion to Lovelace:

> We all know what an inventive genius thou art master of: we are all sensible that thou hast *a head to contrive, and a heart to execute.* Have I not called thine *the plottingest heart in the universe?* I called it so upon knowledge. What wouldst thou *more?* Why should it be the most *villainous,* as well as the most *able?* Marry the lady; and *when* married, let her know what a number of contrivances thou hadst in readiness to play off. (Richardson 1932: 2.319)

Lovelace need not realize his contrivances: it is enough for him to offer a modal account of them. He can tell the true story of what he *could* have done, which is an honest signal of his powers because he alone could have thought of it.[1] His power over her is like the authoritative power that purveyors of plot or contrivance or narrative have over the characters and the stories they tell. I don't mean to make the obvious (and true) point that Lovelace is a surrogate for Richardson the novelist here, but rather that Belford is one surrogate for a narratee, begging the contriver of plot for a different outcome, and that Clarissa is another, but most of all that the novelist or quasi-narrator becomes a kind of surrogate for Lovelace. By this I mean that we are anxious about what the storyteller will do, how he will contrive his own plot, what he will *allow* Lovelace to do. We are anxious about what Lovelace will do, and therefore anxious about what Richardson will do, how he will *respond* to or reciprocate Lovelace's plotting. He controls Lovelace, yes, but we are interested in what he does only because we are interested in Lovelace, and we regard what he makes Lovelace do as a reaction to Lovelace himself. (It's in this sense that he's Lovelace's surrogate.) An author's plots punish or reward a character. The character has a quasi-autonomous existence within the story that the author contrives for or against him or her.

Every story tells another story, the story of its contriver's decisions about what will happen in it. That contriver can be an author, or a studio, or a network, or chance, or fate, or God, or history. When we track what happens in a narrative we track the contriver's choices and narrative decisions as well. This is the outermost in the series of nested dramas in which we take vicarious interest. We are anxious to know how the interaction between narrator and narratee or author and reader will come out.

There are many constraints on whether and how an author can gratify an audience's desires. Perhaps the primary one, which I mentioned above, is that of the verisimilitudinous or plausible. (Such constraints provide the parameters for the human experiments in Zola's "experimental novels"; his novels are experiments in plausibility, since his characters have to act plausibly in conditions dictated by inherited traits and environment.) We are entitled to judge whether a moment or scene or character or event or plot is plausible, but we can't change the plot ourselves (although in the retelling of oral tradition or tales we can, but only if we assure our auditors that we are telling the story as we heard it). The narrator or contriver of a work has authority over what happens, and that is because the narrator or contriver is part of the story, not just its conduit. This is why we can't simply prefer our own fictional outcomes, but instead have to petition Dickens or J. K. Rowling for the outcome we want.[2] It should be surprising that fan fiction, no matter how good it is, is always secondary and not part of the "*real* fictional world" determined by the original contriver. But this is because the author is him- or herself an altruistic punisher and rewarder within the story and therefore part of the story, not just its reporter. The constraints of verisimilitude are part of this larger plot: the author or narrator is also responsible for giving *us* a plausible story, as part of his or her responsibility to us.[3]

We think of narrators as rewarding or punishing by the plots they devise and by the outcomes of those plots. This suggests that our sense of basic character comes first: we are on the side of certain characters and against others, and we wish the narrator to share and effect our desires. The narrator is therefore one character among others, a character who can determine what happens to the other characters, but not really (in our most basic responses) an agency who creates those other characters. The narration of plot or story therefore does two things. First, it reports on the doings of various characters, where the report itself is an aspect of reward or punishment since it opens those characters to the approval or censure of those who attend to the plot. Second, it is the narrator's device for rewarding or punishing by fiat, that is, by plot-contrivance.

As we've seen, the first aspect of plot—report and establishment of good or ill reputation—is not itself enough to satisfy our sense of justice or injustice in a story. The fact that Shakespeare can make it possible for *us* to know (through his unheard soliloquy) that Claudius is guilty of King Hamlet's murder doesn't expose him to punishment within the fiction to which he belongs. We want him exposed to other characters in his world. Villains get away with it if only the reader knows the truth. If only the reader knows who done it, no one knows, just as no one sees what Arthur Conan Doyle (1902) nevertheless describes as a sight: "It was a strange sight, had there been anything but the buzzards to see it" (*A Study in Scarlet,* p. 122).[4] We may hate the villain, but our hatred is meaningless. We want him unmasked to people in his world.

It's this second desire that the narrator can fulfill. We don't just want the narrator to tell *us* who the criminal is (though this can satisfy our curiosity or interest in puzzles or even in hypocrisy when we try to puzzle it out); we want the narrator to make the criminal known to those who can punish him, even if it's just by their knowledge. We'll be well-disposed to narrators who arrange suitable punishment and reward, and ill-disposed to those who refuse to do so.

We can therefore say that in cases where the narrators arrange for some character to engage in altruistic punishment or other genuinely altruistic activity they operate at a threshold between the world of the fiction or narrative and the world of the audience or reader. The narrator takes on the task of making sure that some character punishes or otherwise strongly reciprocates the actions of other characters in the fiction. We approve or censure *narrators* as a function of whether and how they do this—how well their narratives end up enacting and executing our approval or censure of characters. Their altruistic stance is directed "inward," toward the fictional world, as well as "outward," toward the social group to whom they offer the story.

Narrators solve narrative problems for us, and it's no accident that we respond with a quasi-moralistic vocabulary of critical praise or blame: a good story, a great movie, a ridiculous plot twist, a disappointing ending, a story that cheats. Although I'd certainly concede that these are judgments of "taste," they are to be analogized more to our taste for certain kinds of behavior than for certain kinds of food.

I want to look at a few ways that narrators and authors solve these problems for us. For simplicity's sake I won't here make much of a distinction between narrators and authors, as I won't make a distinction between narratees and readers or spectators, having taken some pains to analyze these distinctions earlier. Suffice it to say here that they contribute to the levels of monitoring and strong reciprocity in which we engage. I approve of an author's arranging things so that a narrator can gratify a narratee through the way a character treats other characters altruistically and attains approval or censure within the world of the fiction. My strong reciprocity, my vicarious experience, therefore extends to 1) the characters in the fiction; 2) the quasi-fictional implied community constituted by the fact that the story is being told, especially narrator and narratee; and 3) other members of a real audience to the story, and the real purveyors of that story.

Successful narrators can do the various things that I have argued are essential to strong reciprocity, and we will therefore approve of their ability to do these things in approving of their strong reciprocity, of the way they'll solve narrative problems. This is why we also approve of their skill in solving narrative problems they invent, and we are particularly prone to admire them when their solutions are better than our own might be. (Riddlers impress us because they can answer their riddles for us, not because they can pose them, as poor Stephen Dedalus finds out, perhaps, at the end of the second chapter of Joyce's *Ulysses* [1986]). I would like therefore to look at a set of examples of narrators engaging in the components that together comprise the context of strong reciprocity. These components include, in particular: monitoring, signaling, and arranging for punishment and reward within a narrative.

In some narratives more stress is laid on one aspect, in others on another. In very few works of literature does the narrator or author need to take on the burden of sustaining all three at a very high level. Early Shakespearean comedy might do so, for example *The Comedy of Errors* (2001), but later Shakespeare saw no need to do this. The audience is capable of cooperating with the narrator here. Generally or generically, monitoring plot is the aspect the audience has least difficulty with. It's not hard to remember that the stepmother wants to leave Hansel and Gretel in the forest, but that she doesn't know they suspect it. But in *The Comedy of Errors,* Shakespeare (or the play, or the fictional state of affairs) always remembers the complex situation better than we do, and we take delight at every moment of apt reminder of the situation and how it has developed.

Narrators (like joke-tellers) signal narrative intention, and to the extent that surprise or narrative resourcefulness is part of the pleasure of narrative, narrators will challenge audiences by overtly signaling some plot elements in advance—elements they can afford to telegraph. Again, we can know that it's not Little Red Riding Hood's grandmother in the bed, but the wolf, and we can relish (or fear) the moment of recognition that is to come. We participate by anticipating, and good stories (like good jokes) can afford to let us do so. Sometimes we're right and sometimes we're wrong, but in both cases the storyteller had better have a good trick up his or her sleeve. It's not necessary to telegraph plot points much, but when a story does, it must keep something in reserve to make the story worth the candle. What will happen *after* Red Riding Hood recognizes the wolf? That's what the story tells us only in its own good time. Again, the *Comedy of Errors* signals some of the recognitions and reunions to come, and we wait for them with grateful expectation, but the real surprise is the reunion we have no idea is coming. The general principle here might be taken from Hitchcock: "Tell the audience what you're going to do and make them wonder how." The signal *what* is a costly one, since it gives up some of the element of surprise. But the surprise is interesting, and available to the audience, only to the extent that they're *looking out for* it: "In the theater the audience wants to be surprised but by things that they expect," as Tristan Bernard put it.[5]

It sounds almost trivial but is, in fact, fundamental to say that narrators demonstrate their superior ownership of the stories they tell by signaling where they will go. Such signals might most frequently take the form of announcements of genre and of other stereotypical markers. The storyteller bets on being able to surprise an audience despite telling it what to expect, an audience with whom he or she is as much in competition as in setting riddles or telling jokes. (It's no accident that so many stories revolve around riddles and their solutions.) Now, remember that storytelling itself is an altruistic act, whether as display or as reward or punishment or both. Storytelling may be a species of a more general propensity to altruism. It may be altruistic without itself necessarily functioning as a *display* of altruism. But to the extent that it does function as such a display, then there are three aspects of altruism a storyteller might display through the story. The first—the care lavished on producing something for the audience—would have to do with the particularly aesthetic qualities of the work, and is, of course, widely studied in accounts of the sources of people's consuming commitments to their work. I have nothing particular to add here: it seems uncontroversially true that art is a costly signal of underlying talent, commitment, or vocation, and that these are all positive items in a person's reputation. I'll return to just one component of this later on, which has to do with the display of the author's cultural capital, whether in a social mode (James, Wharton, Woolf, Proust, Thomas Harris, Robert Ludlum) or in an intellectual one (Joyce, George Eliot) or in the mode of expertise (Melville, Arthur Conan Doyle, Philip Roth, Patricia Cornwall, Kathy Reichs, Tom Clancy, Ludlum again, Cormac McCarthy).

More relevant to my argument are the narrative capacities storytelling may signal about the teller, in particular superior ability to monitor, and superior ability to arrange an appropriate ending. Again, *The Comedy of Errors* monitors and keeps track of everything and then arranges the happy ending, including one that we're not expecting. Shakespeare gives us what we want by giving the characters what they want; more than that, he gives us more than we even had the wit to want by including Egeon and Emelia in the happy ending.

I will give a few examples of cases where the narrative focuses on one or another of these basic elements of the storyteller's superior command of the story. We can start, as we started the book, with the question of monitoring, or how narrators keep track of things.

Monitoring

Something that impresses me in both Homer and Philip Roth is their ability to keep track of what's happening in their narratives. The *Odyssey* especially is dazzling in the way Homer can keep nested stories separate and consistent, and never lose his place within whatever nest he's in. (Interestingly, this is more of an aspect of *The Odyssey* than of the infinitely careful *Ulysses*, where the transitions from chapter to chapter are intentionally discontinuous.) Roth too likes to produce a surprising consistency in his novels. After some minor scene-setting, he'll go off on a hundred-page sequence, and the essayistic or polemical or analytical aspect of the novel will crowd out the almost perfunctory set-up. But then, when the reader has completely forgotten it, that set-up will return, unimportant in itself but a scene or passage to other events, and it will transpire that the plot itself was far more important, and far more consistently arranged, than we had any idea. Proust (1987) does this too, but much more overtly: we find out, for example, that Gilberte was walking with Lea and not with a young man two thousand pages earlier. Paradoxically enough, in Proust the effect militates against a convincing imitation of memory, since the moments whose meaning is revealed much later are set-ups for the revelations, and thus turn out to be high and overt artifice. In Roth, though, the sense he manages to give is of genuine memory. Zuckerman's memory of the initial scene was perfunctory but *real,* so that he remembers it even if we don't, not because of its significance but because of its (fictional) reality. It may be idiosyncratic of me to marvel at Roth's ability to do this apparently simple thing, but I do, and wish to offer it as one example of what I expect is a more general delight in this kind of thing.

A narrator's monitoring can apply to two different things, to the characters, incidents, and other elements of the history he or she is recounting, and to the discourse by which the narrator recounts that history. If Roth and Homer are really good at keeping track of what their story says has happened, at remembering recounted events, mystery writers have to be good at monitoring their own discourse, keeping track of how they tell their stories. We think of a whodunit as not cheating when it gives you all the information you would need to solve the crime, without telling you anything that would prevent you from solving it, but still manages to mislead. The trick is to present the ambiguous moments as though they're not ambiguous. The narrator has to sustain awareness of how an audience is interpreting what he or she says, say things consistent with what he or she knows the audience's interpretation to be, but at the same time tell a truth that goes counter to that interpretation. In a successful mystery, the narrator can keep track of our responses or reactions to what he or she is saying even as the narrator also keeps track of what a proper or demystified response would be. This is obvious enough not to need analysis. Agatha Christie's novel *The Murder of Roger Akroyd* (1983), or the subtle turn on that novel that Scott Turow gives in his first novel, *Presumed Innocent* (1987), are classics of accurate misdirection.

It's more intriguing when such ambiguity occurs as an element of plot rather than as an element of presentation. James's *The Turn of the Screw* (1966) can serve as a transitional example between the authorial misdirection or ambiguity in presentation I've just noted and intentional ambiguity as an element of plot. Let's stipulate that the novel is really two different stories telescoped together: real ghosts versus governess's hallucination. The point then is to see not how James misdirects us as much as how every plot element in *either* story (even taken singly) is or can be willfully ambiguous, and how a good reader understands this quasi-willful ambiguity among James's characters. The governess behaves in such a way that the children will know she knows they are consorting with the ghosts *if* they are consorting with them, but will suspect nothing if they are not; they too may be testing the governess's understanding of their motives by seeming to acknowledge them frankly if she does understand, while sustaining plausibly innocent behavior if she doesn't. James's later heroes and heroines are skilled in the uses of ambiguity, including ambiguity about whether they are being ambiguous. The pleasure of tracking them is in seeing how they use it, not in being fooled by how they use it, as it is in a mystery. But James sees this better than we do, and always has to explain it to us, to get us to catch up to what he's able to keep track of. What he can and does keep track of culminates in Maggie's hyperartful utterances, which signal three different things depending on what the three people who hear them know or might know and what they think or might think she knows.[6]

Let's consider another example in a little more detail: the dazzling end of Dashiell Hammet's story "$106,000 Blood Money," the sequel to "The Big Knockover" (Hammett 1989; Hammett loved Henry James). At the end of the story, the Continental Operative, who narrates it, has to manage the interactions between a dizzying array of characters. ("Wheels within wheels," he says.) He has made a deal with a sadistic but trustworthy criminal, Tom-Tom Carey, to find the hide-out of the principals behind the bank robberies that give "The Big Knockover" its title—Papadapoulis and Big Flora. The Op knows—but none of his allies do—that another operative, Jack Hounihan, is actually working for Flora and Papadapoulis. Not only does Tom-Tom not know this; he doesn't even know that Jack is an operative. He does know that Mickey Linehan and Andy MacElroy are. So the situation in the climactic scene is this: Papadapoulis and Flora are waiting for Jack to double-cross the Op and Tom-Tom, so they can kill them and escape. They don't know that the Op is on to them. The Op knows that Mickey and Andy think that Jack is on the level. He knows that Tom-Tom doesn't know who Jack is. He knows—most crucially—that Mickey and Andy don't know that Tom-Tom doesn't know who Jack is. Because the Op knows who Jack is, he doesn't let him get the drop on himself and Tom-Tom, and Tom-Tom kills Papadapoulis, But things aren't over yet: now the situation is, he tells us, as he wanted it. He tells Tom-Tom to watch Jack carefully. Then, the Op arranges a private scene in which he accuses Jack of having gone wrong, a scene he gets Tom-Tom to watch from a distance. Jack raises his gun to shoot the Op, but Tom-Tom reacts swiftly and kills Jack to save the Op's life, as the Op knew he would. Tom-Tom does this because all he sees is a man whose relation to the Op he doesn't know trying to shoot him. But what Mickey and Andy see is Tom-Tom shooting an operative (one they do not know has betrayed them), so they shoot and kill Tom-Tom. The Op is conscience-stung by what he's done, but manages to maintain a straight face. The Op knows that Big Flora knows that Jack is an operative, although corrupt, and so she alone understands what has happened. But the Op manages to silence her by comparing his own Machiavellian behavior to that of her beloved Papadapoulis.[7]

This barebones summary of a fast and powerful scene should help make the point that what the Op is better at doing than anyone else is keeping track of everyone's relation to everyone else. He doesn't slip up. He knows the extent and limitations of everyone's knowledge of everyone else: in particular that Tom-Tom doesn't know that Jack is an operative, and that Foley and Linehan don't know that Tom-Tom doesn't know this, and also that Jack has forgotten that Tom-Tom doesn't know this. The Continental Op controls the scene because he has the monitoring talent of the strong reciprocator or second-order altruistic punisher. He regulates the punishing others undertake: in this case Tom-Tom's of Papadapoulis (for the reward) and of Jack (altruistic), and Mickey and Andy's of Tom-Tom (altruistic). He is not doing this for the reward (it is well-established), so that what impresses us about the Op is just his prodigious talent as an observant strong reciprocator. (The story ends with his explanation of his play to his boss, the fathomless Old Man. But now, for the first time, he knows what the Old Man is thinking as well, and so he has acceded to equality with the Continental Detective Agency's dominant or alpha male.)

It would be enough to note that he is the narrator of the story to show the extent to which the narrator is too fast for us and impresses us with his superior powers of keeping track of what others are able to keep track of, in order to play off those he keeps track of. We admire altruists and monitor the component talents of a candidate for our admiration, and therefore we admire the Op for his monitoring capacities (and then the use he puts them to as a punisher or regulator of punishment). This wouldn't be true if we're thinking of Hammett and not the Op as the storyteller, but even so the fact is that the teller of the tale is able not only to keep track of the different perspectives of all the characters, but of the reader's perspective on these different perspectives as well, and to sustain a perspicacious relation to all of them, his perspective containing ours. The Op and Hammett are both able to make accurate sense of what everyone else is doing, and this alone is enough to command our admiration.

Signaling

This account of monitoring is also, of course, an account of signaling, and especially of the differential signaling that I considered in the Chapter 2. Henry James, Dashiell Hammett, and the Continental Op all keep track of what's going on around them, particularly what and how much others are keeping track of. And this includes what their audience knows and is keeping track of as well, whether that audience is Andy MacElroy and Tom-Tom Carey, or the readers of the book.

But the studied ambiguity that such monitors cultivate (and that, as we have seen, we could find in a gamut of writers from Susanna Moore to Proust to Woolf to Agatha Christie) is not itself a *costly* signal. The revelation of the ambiguity may be, "Look how well I've managed this." But the ambiguity itself is not. To what extent can we think of narrators as engaging in costly signals or honest handicaps *during* the act of narration, and not as a consequence of having narrated?

Think of storytelling as being like a magician's trick. The magician handicaps him- or herself in various ways order to make the trick harder. He or she might be straight-jacketed, or on a tight-rope, or bare-armed. The pungency of the trick is in the combination of handicap during performance and beauty of result. The magician sits on a unicycle (handicap) and makes a dove appear (result). So too the storyteller handicaps him- or herself in telling the story, making the result more difficult, and then arranges a result that will, in one way or another, gratify the audience. One of the elements of a Hollywood screenplay is called the "lost point," around three-quarters of the way through the story, where the protagonist's plans have failed and it looks as though there's no way to achieve success. That depiction of failure is the handicap that makes final success all the more impressive. We don't admire Superman for defeating Lex Luthor unless he has first been caught in an apparently inescapable trap.

You could describe this as a narrative setting a trap for its protagonist and also for itself. But more specifically, not only narratives but also sometimes narrators set these traps for themselves, produce lost points where it looks like they will not be able to get out of the traps of their own making. A quick version of this combination is the wonderful paragraph at the end of *Northanger Abbey* where all is resolved. General Tilney, we are sorry to learn, has refused his consent to the marriage of Catherine and Henry:

> The anxiety, which in this state of their attachment must be the portion of Henry and Catherine, and of all who loved either, as to its final event, can hardly extend, I fear, to the bosom of my readers, who will see in the tell-tale compression of the pages before them, that we are all hastening together to perfect felicity. The means by which their early marriage was effected can be the only doubt: what probable circumstance could work upon a temper like the general's? The circumstance which chiefly availed was the marriage of his daughter with a man of fortune and consequence, which took place in the course of the summer—an accession of dignity that threw him into a fit of good humour, from which he did not recover till after Eleanor had obtained his forgiveness of Henry, and his permission for him "to be a fool if he liked it!" (Austen 1996: 217)

The trick or joke here is in the self-referentiality. The reader is anxious because there is so little time to resolve the plot problem. How can Austen escape this trap? But she turns the problem itself into its solution: with so few pages left, their paucity itself tells the tale she does not need to tell. She pulls a rabbit out of a hat, impossibly and convincingly, where we would not have been able to do so. The narrator can represent herself as utterly insouciant about the costly narrative corner she's painted herself into because it allows her to display her near incredible ability to economize to the point of magic with the amount of space left her. Freud called such economies the heart of jokes, and said they made us happy in our lives. I think they do, and I think the reason they do so is that the teller's grace and nonchalant generosity with time, set-up, anticipation, and even cooperation in understanding *what* tale the compression of pages must tell is something we are made happy to contemplate, lovely narrative largesse we take delight in admiring.

A longer version of such costly signaling's occurring purely on the level of narrative may be found in Cormac McCarthy's *Cities of the Plain* (1998). McCarthy's bravura narration of a knife fight (pp. 246–254) is one of the climactic scenes in that book. John Grady Cole finds himself about to have a battle to the death with the elegant and murderous pimp Eduardo. They are unevenly matched, Eduardo a debonair expert with the blade and John Grady barely past adolescence. Eduardo taunts him as they circle each other, and the taunting functions in an extraordinary way in the narrative. For Eduardo keeps suggesting fantasy scenarios by which John Grady might somehow win.

At one point he offers to sell John Grady back the knife he has lost for the price of an eye:

> If you let me pry one eye from your head I will give you your knife, he said. Otherwise I will simply cut your throat.
>
> The boy said nothing. He watched.
>
> Think about it, said Eduardo. With one eye in your head you still might kill me. A careless slip. A lucky thrust. Who knows? Anything is possible. What do you say? (McCarthy 1998: 252)

Each of these suggestions, each of these hypotheticals—"Anything is possible"—is made impossible through its utterance. A basic rule of competent narrative is that any spelled-out plan, or possible outcome, will go awry. What is narrated as a plan will not be narrated again as an event. We know John Grady will not be saved by a lucky hit, because Eduardo has already contemplated and articulated what we might have hoped would be articulated later by McCarthy instead. In the fight, we imagine possible scenarios by which John Grady might win. We're thinking, "A lucky hit would do it. A careless thrust. Who knows?" Who knows? McCarthy does. Just as Eduardo knows in relation to John

Grady, McCarthy knows in relation to us. Eduardo and McCarthy both make impossible the plausibilities they name, just by naming them.

It will be seen that McCarthy's narrative is itself parrying any anticipatory thrust that we might make. We're in competition with the narrator, looking to be able to solve the riddle, anticipate the happy ending before he gets there. Every possible outcome that McCarthy mentions comes at a cost, both to him and to us: that outcome will not occur. "What does he see," Eduardo taunts John Grady. "Does he still hope for some miracle?" So there can be no miracle. But McCarthy is able to bear the cost of even that narrative constraint, because he has more in reserve, something that we haven't thought of.

It's important to see that here he differs somewhat from Eduardo, who is narrowing possibilities down to the inevitability of Eduardo's triumph and John Grady's death. Where McCarthy is implicitly claiming to be a narrator superior to the reader, claiming (like any writer of genuine suspense) to be able to pull off a surprise that the reader could not pull off, Eduardo is signaling instead his ability to bear the costs of the fight. There are no eventualities that he has not anticipated. Both Eduardo and McCarthy read their auditors' minds. But Eduardo lives, as it were, in a real world, in what is for him *the* real world, and in the real world anticipation doesn't preclude fulfillment. He could be killed by a careless slip or a lucky thrust. In saying so, he signals John Grady his capacity to bear the costs of those risks—his disdain for them. He offers to cooperate with John Grady, to give him the knife back, in exchange for an eye (a man whose eyes have been gouged out, like Oedipus or Gloucester, has been an important figure earlier in the Border Trilogy, and this passage is a thematic recapitulation of that earlier episode). He gives up his absolute advantage. And when John Grady doesn't respond, he taunts him by asking for an ear merely, like a matador already taking a trophy from the bull. John Grady lunges for the knife—we think this might be the narrative moment we'd been hoping for!—but Eduardo is too fast and John Grady backs away, badly wounded. And then Eduardo tells him to pick up the knife: "Pick it up. Did you think I was serious? Pick it up."

This is a grim example of predator-prey cooperation through costly signaling. Why does Eduardo gives John Grady a chance? He is satanic and we don't much care about his motives or anthropomorphize him. His authority is too great. But we can still recognize that he is doing this for John Grady, to make him feel the totality of his abjection and defeat.

Taunting him isn't a means to the end of killing him; killing him is rather the way he makes the taunt absolute: "Perhaps he will see the truth at last in his own intestines," he says after cutting John Grady's stomach.[8]

Eduardo gives John Grady the knife back as a way of proving his absolute superiority to John Grady, a superiority that will even bear the risks of the cliché by which the bad guy makes the mistake of inventing a baroque but imperfect scheme to kill the good guy, so confident is Eduardo that he can bear the risks of overconfidence. Like a matador he turns his back on John Grady, knowing that nothing will happen.

The parry and thrust that have been going on between Eduardo and John Grady, and in tandem between the reader and McCarthy, who expertly parries our hopeful thrusts at a narrative way out, are now complicated by the interesting fact that it is McCarthy who has to give John Grady a way out of the narrative trap, that it is McCarthy who has to solve the problem of defeating Eduardo despite the fact that Eduardo accurately signals his ability to anticipate all contingencies with which we might attack him. And McCarthy does solve the problem, by having John Grady just stop parrying. Eduardo depends on John Grady's defending himself. He attacks, and when John Grady blocks he changes the angle of attack and cuts before John Grady sees where his knife is going. This is called feinting, but the feints turn into direct attacks when the opponent doesn't move to block them. The feint is, in fact, an honest signal of where the attacker might attack; it's only when the opponent blocks that the attacker responds to the block by changing the attack.[9] The attacker *wants* the opponent to block, because in blocking the opponent is defending, not attacking. But John Grady stops blocking. Eduardo, accordingly, stops feinting, and cuts him to ribbons. As he does, John Grady slashes back. Because he is not blocking he can wound even as he is being cut, and he drives his knife through Eduardo's jaw into his brain.

I cite this passage because it overtly thematizes the narrative competition between author and reader in the parallel combat between Eduardo and John Grady. Indeed, Eduardo writes. He carves his initial on John Grady's thigh in three horizontal slashes joined by a vertical. Some such metaphorical relation between cutting (incision) and writing is widespread in literature, from Aeschylus to accounts of the stigmata of Christ and of the saints, to the tattoos of Melville's *Moby Dick* (1967) and the epic that Ishmael plans to write on the foolscap of his own body, to the inscribing device in Kafka's *In the Penal Colony* (1983).

Nor is McCarthy the first to analogize the relation between narrator and auditor of the narrative of the fight to the fight itself. But one might still ask, with justifiable skepticism, whether such an analogy is at all pertinent: whether the kinds of things that go on in narrative expose anyone to anything like the literally visceral danger of a knife fight. McCarthy may talk a good fight, but is there anything in such narrative that goes beyond that?

This is the question that Michel Leiris (1938 and 1946) asks in his two great essays on literature considered as a bullfight. Leiris wants to know whether writing can actually be a dangerous activity, whether the writer exposes him- or herself to a real danger comparable to what the bull's horn threatens the matador with. Are the intense narratives of transgression and pain in the work of Leiris and his quasi-colleagues, such as Bataille and Caillois in the notional "Collège de Sociologie," anything more than metaphorical? To what extent are the costs of writing as tremendous as the costs that writing narrates? Is the analogy melodramatic?

For Leiris and his circle it is almost the only possible sign of a writer's authenticity that his writing should expose him or her to real distress, loss, and sorrow, and it is certainly the case that much literature comprises documents of real distress, loss, and sorrow.[10] I agree with this characterization of one major domain of "literary space," and generally I take that question on its own terms. Here I hope to do it even more justice by treating such exposure—the matador's, the writer's—as an example of the general biological imperative to court real danger, to court others by courting danger. Costly signaling is dangerous, and if narrating has an aspect of costly signaling to it, its most agonized modes may be at least continuous with what the bullfighter does.

McCarthy certainly thinks this is so, since every bitter syllable he writes testifies to the toughness of the writer able to bear the costs of the writing, that is to say, able to bear the violence of the visions he depicts. The reader is also asked to bear something of that cost, asked to know the truth. The truth is the violence of the world. We could say here that this sort of truth—in a deep sense a literary one—is the extreme cost humans pay for being human: for cooperating with one another and, in particular, for cooperating in altruistic punishment. Most of the examples I am offering in this book are examples of very basic narrative scenarios and gratifications and of how they work or what they are about. But if there is such a thing as general literary *depth* the arguments that I am making should be able to say something about that depth, where it comes from, and what it means. What depth means for Leiris and for McCarthy is the ravenous extent to which being in the world involves cooperating in violence, not only violence against defectors and against those from other populations (remember that within-group altruism tends to go in tandem with between-group violence), but cooperating in violence against oneself, or members of one's own group. That cooperation (which we considered briefly in rehearsing accounts of cooperation between predator and prey) is an extreme of altruism. For Leiris and McCarthy, writing requires an exposure to truth, therefore to violence, an exposure that their writing thematizes, since this exposure is the truth they are telling, and the truth is a violent one. For them, to tell the truth is to absorb the fact that the truth is violence. Because the truth is violence their writing takes the narration of violence as itself an exposure to the violence of truth. (This is a special, intense, and, for me in certain moods, uniquely deep case of the more general claim that I am making, which is that attitudes toward narrators and the works they narrate will tend through the logic of strong reciprocity to mirror the attitudes the narrators take toward the characters that figure in their works.)

Leiris relates writing to bullfighting like this. The maneuver by which the bullfighter pivots out of the way (and presents himself in profile) is a gesture by which he renders himself two-dimensional. The bull sees him only edge-on. The matador protects himself from the edge of the bull's horns by becoming an edge himself. In order not to be cut, he becomes a cut himself, a cut in the region or space of the bullfight. Pivoting into profile rotates the plane defined by his body into a cut in the space the plane had represented.

The pivot is risky, since it aligns the bullfighter's body with the bull's sharp and cutting rush down a very close parallel plane. The matador's arch as the bull charges past is meant to minimize the distance between the two edges, to make them vanishingly close to contiguous. The matador's body is very nearly indistinguishable from the cut the bull would make in it. Because he has turned in profile, any such cut would be raking, and if the matador presents a pure edge the limits of the raking cut is the surface of the body itself; the surface and the cut into it are the same.

The matador normally survives the pass. But the pass focuses intention on the vulnerability of his body, on its iconic resemblance to the cut that would destroy it. It's not only that the risk is real; the risk also *rep-*

resents the fragility and frangibility of the body, the fact that human interiority is protected only by surfaces that cannot resist the cut that will lay the interior open to the surface or make of the interior an exposed superficies. The danger is that of exposure, the exposure that makes what is hidden into a surface.

The effect of presenting oneself in profile raises complex questions, since the profile of a face is not an edge but a surface. It's nevertheless a surface that suggests an edge, as on coins and medals where profiles suggest the superiority or disdain of the face seen in profile. She need not and does not look at the viewer. She looks away. And she need not hide anything from the observer. She reduces herself to edge and surface, looking along, past, and with the edge that constitutes her two-dimensionality. She is without interiority and does not look back. Profiles (of adults) always schematize superior power or status because of the way they present everything and are contemptuous of any protected interior. Think of Sartoris Snopes's perception of his father's fearful power as metaphorized by his looking as flat as a tin cut-out in Faulkner's "Barn Burning" (1950). The flatness of the profile represents indifference to any unseen observer's scrutiny, and therefore profiles are strong handicaps or costly signals.

Profiles are easy to cut unless seen edge-on, when they risk being the cut itself as well as cutting you. A glancing relation to the murderousness of the bull's horn is what Leiris aspires to in the literary work. It's through a similarly glancing, raking, oblique relation to the work of art that the writer would imagine entering into its world and being made vulnerable to its violence. The writer puts him- or herself on the page; the artist renders him- or herself into the two-dimensional world of the image. To enter an image you have to turn sideways. The knife-edge sense of the mimetic effraction of and into two-dimensionality has never been better described than by Leiris—the real psychic costs of literary vocation, of experiencing the flow of every passion and intention into the pure exposure of the surface of the work. The work cuts its readers because it is cut itself, because it is a pure cut as well. It is all surface and therefore its edge has the perfect sharpness of the one-dimensional. What it is exposed to is the truth; what it cuts with is pure exposure to truth.

I have risked using the romantic-theoretical language of the Collège de Sociologie and its colleagues—to which I am committed—because I want to insist that I am not *reducing* literary or aesthetic phenomena to ornate examples of biological processes. Evolutionary psychology does much too much of that, accounting for everything about the work of art except the intensity of its experience, which it then more or less explains away or trivializes. Trivialization of subjective experience is a sure sign that the theory that would explain it is incomplete or wrong. But my claim is that McCarthy's relation to what he narrates makes the episode of the knife fight very serious indeed. This sense of the literary is not the phenomenon that I set out to explain in this book, but it derives from the things I do explain—namely from the fact that the costly signaling that occurs in narrative is rightly called costly because those costs are real and serious and may become close to unbearable.

I suspect McCarthy, who quotes German Romantic literature elsewhere, knows Kleist's story-essay on the marionette theater. I think this scene descends from the bear-fencing scene in that story.[11] The narrator recounts his bout fencing a chained bear. The bear will never respond to a feint and never fail to parry a thrust. (He never thrusts himself.) He knows his human antagonist's intention before the human does. The human loses the bout because of the self-consciousness he brings to it. This is one of the three anecdotes in "On the Marionette Theatre" (1810) tending to show that self-consciousness is itself like a wound to the writer. It's not a real wound (the bear never thrusts, the young man who looks like the Spinario sculpture of the boy pulling a thorn out of his foot is only lifting his foot up), but this very deflation is one of the ways that it wounds. The writer is conscious that the wound isn't real, and that he has given everything up, sacrificed his life for a vocation of wounding self-consciousness. It is in this way that writing is like a ritual knife fight, at least from the point of view of the writer. It is in this way that writing is a signal—not an actual fight, but a signal that is genuinely costly and that involves the genuine pain of exposure and the genuine costs of consciousness.

Put differently, one might say that if consciousness or subjectivity is a genuine cost for all of us thrown into the world, then writers can make use of and aggravate the costs of consciousness and subjectivity in their writing, and can display the fact that they are doing so in their writing. They will do so because that's how human beings are constituted: to an altruistic, often painful display of their own ability to bear the costs of being alive. If true altruism nevertheless can give rise to an experience of pleasure, as the brain-scanning evidence suggests, the "indescribable delight" that Gilles Deleuze says always rushes out to us from great books, no matter how grim their subject,[12] might be the analogue of the very costly, bitterly despairing signaling that seems to constitute some of the greatest literary work: "I suffered what I wrote, or viler pain."

Reciprocating

Perhaps it will be a relief to turn to Adam Smith again. One of the moments of vicarious experience that he discusses, in connection with the idea of sympathy and of volunteered affect, is that of being a purveyor of narrative. We've noticed Philip Fisher's (2002) account (citing Hume's parable of the man sleeping in a field oblivious to the danger we see and react to) of how readers volunteer affect. What good storytellers do, in one way or another, is to get us emotionally committed to certain outcomes, and in one way or another to reward us with outcomes that satisfy. Our commitments are voluntary and volunteered, and our emotions may be directed at various characters in the story, or at the storyteller, or both. As I noted above, we'll tend (at least at first blush) to be moved to praise a storyteller who contrives a happy or at least an appropriate ending, and blame one who refuses it. We will also, perhaps in connection to this first attitude, often enter into competition with the storyteller as to who will first guess the solution to the problem that constitutes the story: the reader or the narrator. We could say that the competition is often as to who, the reader or the author, will first declare the way to save the protagonist.

We both have the same goal—the best ending—but in wanting it for the characters we often also compete with each other to provide it first. Since such provision is altruistic in both attitude toward characters and in time and mental energy spent solving their problems (these would include opportunity costs as well, since we could be using out time more profitably) the competition is one between a narrator's altruistic (though not necessarily beneficent) attitude toward us and our attitude toward him or her. In live and improvised narration, the attitudes can go both ways; in reading it's a matter of a reader's strictly profitless satisfaction or self-satisfaction (if the reader wins), or surprise and surprised delight (if the narrator wins). In all these cases the relation between storyteller and reader is what we've been calling cooperative, but not necessarily benign.[13]

But I think that we all feel that most purveyors of narrative are engaged in a more or less benevolent activity. How grateful we are to good storytellers. Smith (2004), in his account of sympathy and volunteered affect, considers the mutuality between storyteller and auditor, and in particular the way the *storyteller* volunteers affect for and sympathizes with the *auditor.* The two are not necessarily distinct, except that the teller is thinking about the auditor, whereas the auditor is thinking about the tale:

> When we have read a book or poem so often that we can no longer find any amusement in reading it by ourselves, we can still take pleasure in reading it to a companion. To him it has all the graces of novelty; we enter into the surprise and admiration which it naturally excites in him, but which it is no longer capable of exciting in us; we consider all the ideas which it presents rather in the light in which they appear to him, than in that to which they appear to ourselves, and we are amused by sympathy with his amusement which thus enlivens our own. On the contrary, we should be vexed if he did not seem to be entertained with it, and we could no longer take any pleasure in reading it to him. (Smith 2004: 8)

This description of the effect of turning into the teller of the story one has heard (a description just as fitting for the social practice of hearing and then telling jokes) seems right to me. We take an interest in imagining and sympathizing with the pleasure or interest someone else would take in a story that gave us pleasure. The idea of inspiration, the idea that the author is not the first originator of a story but repeats it from the Muse, captures something of the vicarious pleasure of imagining the audience's fresh response to a story no longer as fresh to its teller as when he or she first heard it. Acting and, as I've suggested elsewhere (Flesch 2000), teaching will often give the same pleasures.[14] The point is that any storyteller is or has been in the position of being the audience for that very story, which contrariwise allows us to read Smith's point as applicable to original authors as well (Dickens laughing at his own jokes, for example).

To read a pleasurable book or poem to a companion will often mean to take pleasure in our anticipatory knowledge of how what we read will solve the problems that it sets. We feel an altruistic impulse to offer our companion the grace and novelty, surprise and matter for admiration, ideas and amusement the book has once given us, and with these things aim to gratify much or even most of the time our companion's general desire for hearing about strong reciprocity. The storyteller is happy to tell the story in which the characters experience the generous or just outcomes the auditor wants. The storyteller does *for his- or her hearers* what they want him or her to do *for the characters*. It's for both audience and characters, then, that the storyteller arranges things appropriately.

Cringe-Making Scenes

Usually this successful arrangement will be simply a matter of technique and invention. But sometimes storytellers will have to do more, intervene more conspicuously or more obviously, and I want to look at a particularly vivid example of altruistic and even admirable narrative behavior, that of facing up to an awkward and embarrassing narrative situation.

I earlier quoted Smith on one powerful mode of vicarious interest or volunteered affect: that of embarrassment on behalf of someone else. From Malvolio or Falstaff, from Don Quixote to Philip Roth's Mickey Sabbath and most roles played by Jack Nicholson, we squirm at the spectacle of shamelessness. We volunteer shame for the shameless, just because they *are* shameless. Quick examples would include such otherwise different characters as Roger Solmes (Clarissa's awful suitor), Ferdinand Lopez, to whom we'll return, Lapidoth in *Daniel Deronda* (1999), Osborne and Jos Sedley in *Vanity Fair* (1998), Aunt Norris in *Mansfield Park* (2003b), both Emma and Mr. Eliot in *Emma* (2003a), Christopher Newman and also Caspar Goodwood (and any number of Henry James's other young men), both Bloch and the Baron de Charlus in Proust, and so on.

This is a subject that Trollope thought about very deeply and explored in a novel way. In Trollope, as we know, there is a right way to do things. There is also a wrong way, and then there are things that ought not to be done, that a gentleman or a lady does not do. Part of the drama in Trollope arises when someone not a gentleman (Quintus Slide or the egregious Major Pountney) or not a lady (Lady Eustace, for example) puts a lady or a gentleman in a position in which no lady or gentleman would ever wittingly put another. And it is when they do such things that it is difficult to find a way to respond, a way that consists in handling appropriately the wrong-footer—the cad, that is, who has wrong-footed the lady or gentleman.

I take it that D. A. Miller's (1988) brilliant account of Trollope's disciplinary force is essentially this: that (like Edmond Spenser) Trollope fashions a gentleman, or at least fashions propriety, and that such fashioning applies to us readers as much as to his characters. We readers root for George Vavasour and only later do we come to see the virtues of John Gray (in *Can You Forgive Her?* [1973a]), just as (in *The Prime Minister* [1983]) we root for Ferdinand Lopez (seeing him through Emily Wharton's eyes) and only later come to see the virtues of Arthur Fletcher. We come to see their virtues because we come to see just how destructive are the vices of those we had at first admired. It's not, as with Milton's Satan, that their characters are degraded by their author, but that we come to a more mature sense, a more high-church sense, of what is already degraded about their characters. (Perhaps *Mansfield Park*'s [2003b] Mary Crawford would play a similar role in Austen.)

In Platagenet Palliser Trollope meant to fashion a gentleman, as he said, not too like but not wholly unlike Henry Esmond, the slightly priggish gentleman that he so admired Thackeray for inventing. Gentlemanly himself, he says only this of the graces and virtues of the eponymous hero of *The Prime Minister*: "Platagenet Palliser I think to be a very noble gentleman,—such a one as justifies to the nation the seeming anomaly of an hereditary peerage and of primogeniture" (1883: 172). Intended to justify that anomaly, he can do so only if he's convincing, which is all that concerns me here. We can learn more about what might make him convincing with reference to Trollope's remarks on Henry Esmond:

> He is a gentleman from the crown of his head to the sole of his foot. Thackeray had let the whole power of his intellect apply itself to a conception of the character of a gentleman. This man is brave, polished, gifted with that old-fashioned courtesy which ladies used to love, true as steel, loyal as faith himself, with a power of self-abnegation which astonishes the criticising reader when he finds such a virtue carried to such an extent without seeming to be unnatural. To draw the picture of a man and say that he is gifted with all the virtues is easy enough,—easy enough to describe him as performing all the virtues. The difficulty is to put the man on his legs, and make him move about, carrying his virtues with a natural gait, so that the reader shall feel that he is becoming acquainted with flesh and blood, not with a wooden figure. (Trollope 1879: 129)

The difficulty—what astonishes the criticizing reader—is fashioning or producing a gentleman, someone who can handle the ticklish or embarrassing or dreadful situations that others may put him in. In *The Prime Minister* (1983) (perhaps all of Trollope presents variations on this theme), we are interested in how characters will handle themselves when they come to find themselves placed in situations they do not deserve to be placed in, which also means how Trollope will put them on their legs so that they can handle themselves with an easy gait. This is true in all narrative, but what is peculiar to Trollope is the way the characters are called on to handle these circumstances. There is a right way and a wrong way, and the wrong way is always perfectly adequate to handle the binary aspect of the situation. You horsewhip your opponent, or throw him down the stairs, or fight a duel with him. But situations in Trollope are not binary; they occur in a social context, which is to say in the context in which altruism as well as punishment is at issue, and therefore the fundamentally prosocial right way matters.

Part of the right way in Trollope is social courage. The delicacy that prefers silent misery to asking for aid may be virtuous, but it tends to be a fugitive and cloistered virtue, whereas directness is all the more admirable as it requires courage. This is part of what we admire in Lopez, but part also of what makes us ashamed of him. He doesn't shrink from asking for money. He isn't puling or ingratiating. I cringe at the idea that he is going to go back to Mr. Wharton for more money, after swearing that he has enough. I cringe all the more at the thought that he is going to ask Lizzie Eustace to elope with him.

And yet it is not quite as much for Lopez that I cringe as for Trollope himself. How can he bear to write such a scene? He has set himself a near impossible task: to make Lopez consistent with his own dashing audacity even as he humiliates himself. Thus Lopez asks Wharton for money to pay for his Guatemalan venture: "It was now essential that the money for the shares in the San Juan mine should be paid up, and his father-in-law's pocket was still the source from which the enterprising son-in-law had hoped to procure it. Lopez had fully made up his mind to demand it, and thought that the time had now come. And he was resolved that he would not ask it as a favour on bended knee" (Trollope 1983: 2.123–124). Wharton protests:

> "Such consummate impudence I never met in my life before!"
>
> "Nor perhaps so much unprevaricating downright truth. At any rate such is the condition of my affairs. If I am to go the money must be paid this week. I have, perhaps foolishly, put off mentioning the matter till I was sure that I could not raise the sum elsewhere. Though I feel my claim on you to be good, Mr Wharton, it is not pleasant to me to make it."

Wharton is right. and part of what makes him right is that he could never conceive of or contrive such a scene. But Trollope has conceived of it and has contrived it. Trollope has to accept and see through the task of handling Lopez, while we free-riders watch in admiration. It's Trollope who has been wrong-footed. We feel that this scene, so embarrassing for us just to *read,* must be for him a supremely embarrassing one to write. But he does write it. To use his language: the thing has to be done, and so he does it.[15]

The thing I would like to stress here is that the mode of embarrassment isn't quite the embarrassment we feel for a Charlus or a Sabbath or a Basil Ransome (in James's *The Bostonians,* 1956). In those cases the narrator or author shares in the eye-rolling pleasure, which converts embarrassment to scorn. We enjoy a sense of comic outrage. But Trollope's narrator is more akin to the purveyor or narrative agency of *Lolita* (1958). We know what's coming. Now it's time for Humbert to get his wish (since we already know he'll succeed), time for Humbert and Lolita to have sex for the first time. And the idea that Nabokov has set himself the task of narrating that moment is unendurable. We cringe at the thought of his assigning himself the task of writing Humbert's lines, just as we cringe at the thought of Trollope assigning himself the task of writing Lopez's lines (especially after he has married Emily). And we have still to endure his invitation to Lizzie to elope (to be met with her final but not quite accurate judgment: "Mr. Lopez, I think you must be a fool" [Trollope 1983: 141]).

I hope the comparison with *Lolita* is illuminating. The problem is not simply that we are viewing a character whose actions are mortifying. Humbert is not Kinbote, and Lopez is not Major Pountney. They are rather characters whose panache and audacity and even savoir faire we have reason to admire. And it's not that we quite come to discover that they were not what they seemed; it is the very things that make them dashing that also finally make them mortifying (this is Trollope's trademark way of recruiting his readers to *outgrow* a character whom we once admired). Relying on their own endless resourcefulness, they have allowed their admirably insouciant verve to place them in an impossible situation. In this they resemble their authors, who have embraced a scenario in which they come to be confronted with writing such impossible scenes. Trollope may differ from Nabokov only in not showing such obvious sadistic or masochistic relish for the task.

The difference between Trollope and Lopez, though (as between Nabokov and Humbert), is that he *does* have the resourcefulness his character lacks. A gentleman (a term with somewhat different extensions in Trollope and Nabokov, but perhaps the same valence) finds a way out. And this is what we see, not Lopez, but Plantagenet Palliser always do, even when (as in his treatment of Mrs. Finn in *The Duke's Children* [1973b]) he himself has put himself in the wrong. The way out may not be pleasant. It may require considerable self-sacrifice, unfair as all sacrifice is. But the Prime Minister knows not to take the way out Glencora offers, as he knows that he must pay Lopez for his loss, despite the latter's caddishness in demanding the money. And Trollope knows that he must write the scene in which Lopez asks for the money.

Trollope and Nabokov (along with, to some extent, Austen, Thackeray, Eliot, and Edith Wharton) don't quite torment the reader with the embarrassing situations they depict. There's an element of sadism in the kind of narrative that does: the storyteller elicits volunteered embarrassment on the part of the reader, and therefore it's implicit in the story that the narrator likes seeing the narratee squirm. I think Dickens and Hitchcock are very good at that sadistic element, good also at eventually giving us the pleasure of joining with the sadistic point of view that takes pleasure in the discomfiting of the person whose behavior is so inappropriate. (Lizzie's rebuke of Lopez would fit this pattern, although, of course, Trollope has many more cards to play.) But what Trollope tends to do is not to take pleasure from the sadistic contemplation of a character's foolishness, but to relieve those difficulties himself—as he does in *The Duke's Children* (1973b) for perhaps half the major characters, especially Plantagenet and Silverbridge. We feel palpable relief that he has done this for us.

The most humiliating parts of Trollope are those in which someone fails spectacularly to handle arbitrary but necessary social convention, that is, to behave in a prosocial manner. But these parts are a pleasure too, even if we read squinting, because we can see Trollope handling his characters' lapses. Trollope handles Lopez, and Plantagenet Palliser does as well, and so our vicarious interest in the hero of *The Prime Minister* has the same structure as our interest in its purveyor or author. Part of what we always want to know, more or less anxiously, is: what will, what *can*, the narrator do now? We want to know this because the narrator will do it for characters about whom we are concerned, will do it for characters for whom *we* would wish to do something. The narrator is the hero of the social difficulties he intervenes to smooth over or correct, thus intervening through the creation of a heroic character able to do so. The narrator's relation, then, is one of strong reciprocity, handling things for his characters and for us, and it is just this that we have a propensity to monitor and approve.

In the first two chapters I've tried to show how *narrative* appeals to an audience's predilection for monitoring and appropriate reciprocation. We no doubt offer the narratives we know (from "I know that story!" to "I've read *Finnegans Wake!*" to "I love Godard!") as the social capital Pierre Bourdieu (1984) sees elite art as forming; such social capital is a form of costly signaling. In this chapter I've wanted to suggest how *narration* appeals to storytellers by allowing them to display fitness in these particular roles: fitness in monitoring and fitness in reciprocating, and to suggest as well how the very signal of such fitness might involve genuine dangers in which the costs are meant to raise the storyteller to a higher status than that of those who hear the story. In general, these things are not easily separable, nor should they be, and the difference between teller and auditor is not a bright one (since the auditor will repeat the story to others). It is the interaction of these elements, and our evolved disposition to take a pleasurable interest in them, that defines the gratifications that fiction can offer.

4

Vindication and Vindictiveness

My charity is outrage.
—Margaret in Shakespeare's *Richard III*

The Pleasures of Vindication

In rehearsing some of the recent insights of evolutionary biology and behavioral game theory I've been making two related and converging arguments. The argument that has conceptual priority is the one for humans' prosocial disposition. Even when given the opportunity to be free-riders, people tend not to be. We usually cooperate instead of maximizing the net gain we can get out of the cooperation of others. We do this because of an interestingly complex attitude toward others: we have a native inclination toward altruism; we expect others to have such an inclination as well; these inclinations are self-reinforcing since we expect others to pay to punish us if we don't share enough of what we have with them; and therefore our prosocial behavior is partly predicated on an uncalculated expectation that others will behave in accordance with how well we meet or defy their uncalculated expectations for how we behave.

The second argument is just as important, though conceptually it comes after the first. We tend to reward others who engage in genuinely altruistic behavior, and to approve of anyone else who rewards them, and we tend to approve as well of others who punish antisocial or socially noxious behavior. This is a special element in the prosociality the first argument describes. Some such self-monitoring procedure must have evolved to prevent free-riders from outcompeting altruists, and so eventually causing altruism to die out. The more widespread this self-monitoring procedure, the less it costs to be an altruist.

This second argument is more or less where I have been locating our interest in narrative. We care about the narrative report of what some people do to other people because we care about whether they treat them altruistically or selfishly. Remember that *altruistically* needn't mean *generously.* They treat others altruistically by punishing them for defecting as well as by rewarding them for their own altruism. They treat them selfishly by ignoring an obligation to punish defectors, as well as by defecting themselves. We like characters who engage in effective altruistic behavior and we dislike their opposites. Parties to narrative (narrator and narratee, author and reader) are then in at least a third-order relation to the behavior represented in narratives. We could say, in very general terms, that in narrative at least one character undertakes prosocial behavior either on behalf of or against another character (it needn't be but can be both), and that we approve of the prosocial character. Our relation is therefore one of approval of a genuine altruist. We ourselves can't reward or punish the character we want to see rewarded or punished, but we can cheer on the altruistic character who does—and the storyteller who arranges these things as well.

Most important, perhaps, is that we seem to like altruistic punishment even more than we like altruistic reward. In approving, or praising, or otherwise rewarding an altruistic punisher we get the best of both strong-reciprocal attitudes: we applaud someone who punishes—Achilles, Hamlet, Bel-Imperia (in Kyd's *The Spanish Tragedy* [1999]), Emma, Batman. And athwart this double pleasure we may add another fundamentally prosocial pleasure: the triumphantly volunteered discomfiture, the vicarious pleasure or aggressive sympathy we take in contemplating the experience of being punished, as in that pitch-perfect *Onion* headline I cited earlier: *Hijackers Surprised to Find Themselves in Hell.*

I will risk saying, therefore, that we can model the component of pleasure that plot and character yield in narratives by considering representations of vindication. The narrative structure of successful vindication usually entails the following features. There is a witness who acknowledges someone else's vindication. Such witnesses belong to the narrative world and their accurate judgment and recognition of the truth establishes that truth within the narrative world. Opposed to the character vindicated is the character confuted by the vindication. Such a character may be a defector—that is, may be the character who gains advantage through the spread of false and harmful beliefs about the character who will eventually be vindicated. But also confuted will be those who hold those false beliefs, which is to say every bad witness, every witness whose false judgment and misrecognition spreads or stands for false belief in the narrative world. As we'll see, these bad witnesses may include defectors as well, that is, defectors from our point of view, a point of view we take to be correct. We take pleasure in anticipating the confutation of such characters in a story, perhaps more pleasure than their actual confutation yields. (It is the first element of wisdom, and a repeated lesson in Proust, to notice that actually saying "I told you so" is never as satisfying as looking forward to the time when you'll be able to say it.)

Vindication or the fantasy of vindication therefore is a mode of altruistic punishment, since it offers pleasure in confuting and correcting those who have done harm through their misjudgments. The most harmful sort of misjudgment of this sort is *vindictiveness,* where being vindictive means something like altruistic punishment misfiring. Vindictiveness and spite are near synonyms, and it is worth remembering that spite is another name for altruistic punishment. To be spiteful is to be willing to pay a net cost to harm others. But it's a word we tend to use when we think the altruism it indicates has no prosocial benefits—when it doesn't contribute to our sense of justice or fairness. It feels opposed to the sense of vindication we celebrate when it looks like justice and fairness are being promoted. We can sense that opposition in that vindication tends to take the particular form of the altruistic punishment of vindictiveness. What we call justice triumphs over what we call spite.[1] But the triumph of justice (the victory of *δικη*) has the same structure as the vindictiveness it successfully confutes. One group's spite is another's heroism—as Samson's destruction of the Philistines shows, a destruction that might perhaps better be called spite than the altruistic punishment for which the Israelites praise him.

Here are some quick examples of what I mean. In *The Lady Vanishes* (Hitchcock 1998), we know that Miss Froy has indeed vanished. No one but Iris accepts this. Gilbert Redman would like to believe her, but in the face of so much countervailing evidence he can't. Others like Dr. Hartz doubt her too, although it transpires that this is evil deception (or defection), not Gilbert's insensitivity. Caldicott and Charters, the gay cricket fans, also refuse to acknowledge Miss Froy's disappearance, for reasons that equally baffle and vex us; Mr. Todhunter and his supposed wife have more rational, but more distasteful reasons for refusing to confirm Iris's story. We know that Iris will be vindicated, but we don't know how; we know that those who disbelieve her will get their comeuppance.

The situation then has the appropriately named Iris altruistically looking out for the innocent Miss Froy. (She is the innocent in the narrative scheme, despite the Maguffin revelation that she too is a spy.) Iris's altruistic behavior is misinterpreted by all those around her, and it is their misinterpretation that fuels our interest in the story. The reasons for this misinterpretation vary: the bad guys know Iris is right but are gas-lighting her; Gilbert accepts their deceitful interpretation; Caldicott and Charters can't imagine that Miss Froy's disappearance counts for anything against the cricket match they are traveling to; and Todhunter is protecting his interests. Then, in the gratifying denouement, Gilbert finds evidence that Iris is right, just when she has given up hope, and the bad guys or cheaters are exposed. More satisfyingly, perhaps, Caldicott and Charters understand their duty and compensate for their pigheadedness, thoroughly learning the redemptive lesson we wish vindication to teach, even as the punishment we wish vindication to exact on characters who *must therefore remain unredeemed* is suffered by the somewhat vindictive Todhunter. All the good guys cooperate to save Miss Froy, who turns out to be both innocent and an altruistic cooperator, risking all to smuggle out the information that will save the kingdom. There is no one better than Hitchcock at manipulating the audience's sense of vindication and revenge.

I should acknowledge that vindictiveness doesn't play much of a role in most of Hitchcock's movies, whose villains are all very cool and rational (though minions like Leonard in *North by Northwest* [1988] are sometimes vindictive). The major exception to this rule is *Notorious* (1990). There, Devlin is a Gilbert writ very large, and his refusal of sympathy for Alicia is shocking. But he does come around and saves her from the gruesomely vindictive Madame Sebastian, by using their own self-righteous logic against her and her son.

What's true of Hitchcock is true of Shakespeare as well. In *Hamlet* (2001), for example, the vindictive Laertes comes to punish Hamlet for the death of the innocent Ophelia. His willingness to cut Hamlet's throat in the church is what we call spiteful, since he gives up his own claim to salvation by violating sanctuary. But that spite parallels Hamlet's own altruistic punishment—"Heaven hath pleased it so / To punish me with this and this with me / That I must be their scourge and minister"—the punishment that he suffers great costs to exact (Shakespeare, *Hamlet* 2001: III.iv.175–177). Hamlet too fears that he may be putting his soul at risk (perhaps the devil abuses him to damn him), and in choosing to kill Claudius when he is carousing and not praying, in order to assure his damnation, he would violate well-established canon law and commit a mortal sin. Hamlet and Laertes alike pay the costs of scrutiny, of establishing the truth of the situation in which they mean to act as avengers or altruistic punishers. Laertes listens to Claudius's account, and Hamlet moves mountains to confirm the ghost's. Hamlet is vindicated in the end, before a range of figures: the evil Claudius whose perfidy is exposed (but who never harbored false beliefs); the lackeys Rosencrantz and Guildenstern, more or less equivalent to Todhunter in *The Lady Vanishes* (Hitchcock 1998), whose toadying is confuted decisively; the wishful Gertrude who did harbor false beliefs and hoped that Claudius might turn out not to be evil; and, perhaps most important, the converted Laertes who, like Caldicott and Charters, gratifyingly acknowledges his perverse enmity toward Hamlet and does the right thing in the end; Horatio, who has remained skeptical throughout; and Fortinbras, who will learn and publish the final truth of the story. You may disagree about Hamlet and whether Laertes's enmity is in fact perverse; I disagree myself, but I'm schematizing here, not attempting any subtlety, and I will note that both Ophelia and Polonius die without Hamlet's being vindicated to them, which tells very strongly against him.

I argued earlier that it's not enough for an audience to know who the defector is. Someone in the world of the story must know also, in order to establish the truth for the other members of the social group in that world. Likewise it's not enough, generally, for an audience to know that someone is innocent. The very idea of vindication in fiction tends to mean that *we* know that someone is blameless or acting out of prosocial motives that we approve of, but the other members of the social group deny this. An example of such a moment may be found in a couple of peculiarly perverse (that is to say typically Thackerayean) passages in *Vanity Fair* (2003). Just before Waterloo, Osborne comes close to being seduced by Becky. He receives news of his immediate mobilization and contemplates the sleeping Amelia (his neglected wife):

> She had been awake when he first entered her room, but had kept her eyes closed, so that even her wakefulness should not seem to reproach him. But when he had returned, so soon after herself, too, this timid little heart had felt more at ease, and turning towards him as he stepped softly out of the room, she had fallen into a light sleep. George came in

> and looked at her again, entering still more softly. By the pale night-lamp he could see her sweet, pale face—the purple eyelids were fringed and closed, and one round arm, smooth and white, lay outside of the coverlet. Good God, how pure she was; how gentle, how tender, and how friendless! and he, how selfish, brutal, and black with crime! Heart-stained, and shame-stricken, he stood at the bed's foot, and looked at the sleeping girl. How dared he—who was he, to pray for one so spotless! God bless her! God bless her! He came to the bed-side, and looked at the hand, the little soft hand, lying asleep; and he bent over the pillow noiselessly towards the gentle pale face.
>
> Two fair arms closed tenderly round his neck as he stooped down. "I am awake, George," the poor child said, with a sob fit to break the little heart that nestled so closely by his own. She was awake, poor soul, and to what? At that moment a bugle from the Place of Arms began sounding clearly, and was taken up through the town; and amidst the drums of the infantry, and the shrill pipes of the Scotch, the whole city awoke. (Thackeray 2003: 333)

She never finds out George's blessed feeling for her, then; that he bends to her is all that she will be able to know with certainty, and any meaning this gesture might convey is lost first in the sounding of the bugle before their embrace can yield communion ("She was awake, poor soul, and to what?"), and lost in the end to Becky's proof of Osborne's willingness to betray Amelia. Poor Osborne never is vindicated in the book—not with respect to Becky, and not with respect to his father either. Just before going in to see Amelia he writes his father what may possibly be, and in the event is, a farewell letter: "Hope, remorse, ambition, tenderness, and selfish regret filled his heart. He sate down and wrote to his father . . . Dawn faintly streaked the sky as he closed this farewell letter. He sealed it, and kissed the superscription. He thought how he had deserted that generous father, and of the thousand kindnesses which the stern old man had done him" (p. 333). This letter, which we might hope would vindicate him, in fact does not, and it doesn't because our knowledge of Osborne's thoughts and actions doesn't belong to any character in the novel.[2] The moment we promise ourselves at his death, the moment when his father will receive the letter, works out contrary to our expectation and our hope:

> The poor boy's letter did not say much. He had been too proud to acknowledge the tenderness which his heart felt. He only said, that on the

> eve of a great battle, he wished to bid his father farewell, and solemnly to implore his good offices for the wife—it might be for the child—whom he left behind him. He owned with contrition that his irregularities and his extravagance had already wasted a large part of his mother's little fortune. He thanked his father for his former generous conduct; and he promised him that if he fell on the field or survived it, he would act in a manner worthy of the name of George Osborne.
>
> His English habit, pride, awkwardness perhaps, had prevented him from saying more. His father could not see the kiss George had placed on the superscription of his letter. Mr. Osborne dropped it with the bitterest, deadliest pang of balked affection and revenge. His son was still beloved and unforgiven. (pp. 408–409)

Mr. Osborne's revenge is balked as well as his affection, but that is because revenge and affection are two names for almost the same attitude toward another, especially toward a loved one. Neither Osborne nor his father is to be vindicated, but could Mr. Osborne know what we know, he would feel vindicated and forgiving, and could Mr. Osborne know what we know, we would feel that George was beloved and vindicated. And, strangely, it is the final blow to the chances for George's vindication that at the end of the book even we no longer wish to see it, since it might then scotch the happy future proposed for Dobbin and Amelia, who thinks of him only because she no longer feels compelled to be loyal to the dead and (as she thinks) faithless George Osborne.

It is one of the ways that Thackeray ultimately punishes Becky that he doesn't allow any witness to her vindication in the book, to her finally generous behavior at the end. We might notice a similar feature at the end of many of James's novels, put to a fascinating use. In *The Wings of the Dove* (1937), for example, Kate Croy *should* be the witness to Merton's vindication—as Chad *should* be for Strether (a role, however, that Maria Gostrey more or less plays). But like Chad, Kate turns out to be a disappointment. She who understands so much finally doesn't understand Densher, and so there is no one to do so. Milly presumably did, but we are not permitted to observe that scene, and Milly is dead. In *The Golden Bowl* (1987), on the one hand we don't know what Adam Verver knows in the end; on the other hand Amerigo does understand Maggie, is up to the task of seeing her vindicated. Structurally he plays the same role as Ralph Touchett in *Portrait of a Lady* (1999), who is oddly enough a prefiguration of Milly Theale dying.

Such failure to vindicate a character's just or repentant behavior may be contrasted with the more or less reverse scenario of a novel like Conrad's *Heart of Darkness* (1998). Marlow tells the Intended with feigning voice of Kurtz's dying love for her; to spare her he doesn't punish Kurtz, who is perhaps antisociality itself. This would be a far less competent story without the frame in which Marlow tells the story to the guardians of civilization, the Director, the Lawyer, the Accountant, and, as we might capitalize it, the Narrator. He tells them, at least, the truth, so that his narrative has an effect in the judgments of the world he lives in and so establishes the possibility of cooperation.[3]

Sometimes an audience feels vindicated itself, in addition to being vicariously gratified by a character's vindication. In mysteries especially we are not granted omniscience corresponding to that of the narrator, whether standardly omniscient as in the third-person *Maltese Falcon* (1999), or retrospectively so, as in the first-person Continental Op stories, or Chandler's Philip Marlowe series, or in another way, Sherlock Holmes, Miss Marple or Hercule Poirot stories. It may be more or less standard in most nonmystery omniscient narratives that there is nothing that *turns out* to be the case (at least nothing for the narratee, who is assumed by the narrator to have information that a real life audience might not—that the world is run by apes and not humans, for example, or that a war took place in 2116). The narrator keeps no secrets, or at least none not kept by time itself until the proper moment in the sequence of mimetic recountal. The exception proves the rule: when it transpires that Gwendolyn has taken steps to kill Grandcourt, we are amazed. Eliot just doesn't do that sort of thing, doesn't keep information so material, information known to her protagonist, from her readers. In mysteries, however, we often take the accused's innocence on faith, and also we often take the well-foundedness of the detective hero's skepticism about the accused's guilt on faith as well. We relish the moment when this faith will be vindicated, when Holmes will show up the patronizing inspector Athelney Jones and exonerate Thaddeus Sholto, and therefore relish our outrage that Jones should disparage Holmes and accuse Sholto.

We can measure the deep-seatedness of this expectation of narrative from the surprise that violating such a norm can produce, as in Christie's *Witness for the Prosecution* (2002). There the gratifying exoneration turns out to be unjust. But, of course, that's not the end of the story—such reversals rarely are—and we still get to see the defector punished in the end.

Seeing Vengeance

I mention the expectation that our own judgment will be vindicated by the event because this helps to get at a fundamental aspect of plot: the conflict between true and false vindication, that is, the conflict between laudable vindication and the blameworthy vindictiveness it doubles and confounds—the vindictiveness Marías describes of the vicious thug become a celebrated writer, who falsely believes he's doing himself proud by narrating the tale of how he acted the part of a matador and bated and killed a man like a bull in a particularly horrific scene from the Spanish Civil War (2005: 2.262–263). I want to conclude by considering two such narratives as a way to observe how these issues come together: *King Lear* (2001) and *Oliver Twist* (1961).

Throughout this book I've been treating Shakespeare as a frequent reference for the parameters of narrative, not just because he's so good at it but because the nature of vicarious experience seems to have been something he thought a great deal about. I think this is true of Dickens as well. One might, for example, treat *A Christmas Carol* (1996) as a demonstration of the priority of vicarious experience like that which Homer gives when he has Odysseus weep, not when he undergoes the actual experiences, but when he *hears them recounted* by Demodokos (this is also like Nashe's fantasy about Talbot watching the Talbot scenes). He volunteers affect for himself, and so too does Scrooge when he pities the dead person whom no one mourns in the third stave (Christmas Future) of *A Christmas Carol*. Because a willingness to volunteer sympathy is a sign of virtue and goodness in Dickens, Scrooge's vicarious concern for himself shows that humanity and generosity are dawning within him, and within us too who are so glad to see it dawning within him. This is to say that Dickens is particularly committed to the view, often derided as sentimental but nevertheless widespread, that the audience's desire to reward the innocent and refute the guilty is central to our moral interest in fiction.

I don't particularly want to defend this view (although this book may constitute something of an explication of its reasonableness). I want instead to make the interesting converse point: that Dickens is so good a storyteller because he recognizes the all-absorbing qualities of narratives of innocence persecuted and finally vindicated—recognizes that human auditors are likely to feel themselves fully engaged by such stories because that is how we are. We have evolved to be that way.

Oliver, Trusted and Reviled

My first example is from *Oliver Twist* (1961), and has to do with people's judgments as to Oliver's trustworthiness. After Oliver is saved from the false accusation of picking his pocket, the kindly Mr. Brownlow takes him home, gives him a new suit of clothes, and asks him to take, unsupervised, a five-pound note as well as books worth the same amount of money back to a poor bookseller. His spirit is aroused as he gives Oliver this errand, since he wishes to prove Oliver's honesty to his friend Mr. Grimwig and so to show Mr. Grimwig up. Mr. Grimwig doubts that Oliver will return, such a return being contrary to what Mr. Grimwig thinks Oliver will perceive as his own interests, which would be to abscond with the clothes Mr. Brownlow has furnished him, with the money, and with the valuable books. Mr. Brownlow is sure he will return, and that return will vindicate Oliver to Mr. Grimwig and, in conformity to the scheme I have been outlining, vindicate Mr. Brownlow as well: Mr. Brownlow and Mr. Grimwig are the audience of Oliver's action.

Oliver, luckless boy, is on his way to fulfill his errand scrupulously when he is waylaid by Bill Sikes and Nancy, who force him to come with them. They do this by accusing him in the public street of having deserted his family—Nancy claims to be his sister—so that his denials and pleas infuriate the abduction's spectators, who think that Oliver is lying. They condone Nancy and Bill's apprehension of Oliver. Meanwhile, Mr. Brownlow and Mr. Grimwig wait to see whether Oliver will return, and who will win the contest as to accurate prediction of Oliver's behavior and character.

The scene is a highly effective one, and belongs to a highly effective genre, that of a frame-up. (Hitchcock is the great cinematic exponent of this situation.) We're aware of the frame, and in our own response we'll typically anticipate vindication while resenting the vindictiveness of those who don't recognize it. As we've noted, the two things are related. Mr. Brownlow and Mr. Grimwig bait each other: Mr. Grimwig with a "provoking smile," Mr. Brownlow with a "confident" one. The different tonalities of the phrases—*provoking* versus *confident smile*—suggest a difference in our sense of the two old men, but the similarities matter also: each is confident of being right and provoked by the other's confidence. Likewise, vindication and vindictiveness name differences in degree and in the moral judgment of whoever uses one term or the other.

Vindictiveness is spiteful, whereas vindication demonstrates the reasonableness of the vindicated person's altruism or trust or risk. We might add a further distinction by saying that vindication is something like a demonstration that someone has been wronged or judged unfairly or even vindictively, that demonstration being a triumphant end in itself, while vindictiveness has a far greater component of overweening revenge in it (Bacon's wild justice), so that the vindictive seek not only to be vindicated but to harm the adversaries they conceive have harmed them. They imagine that they are entitled to cause great harm—that they would be acquitted for doing so—because their very willingness to show their spite to a world that disapproves of spite would be a sign of how great was the wrong they suffered: Look how spiteful you've made me; look how spiteful she's made me! Spite is perhaps a sign of how wronged they do, in fact, feel. To risk looking vindictive is itself a kind of emotional anticipation of vindication (one almost always disappointed, and sometimes spitefully proud of being disappointed).

Spite isn't particularly evident in this scene (although it's to be found passim in the novel), but the vector by which vindication threatens to turn into vindictiveness is. Mr. Brownlow anticipates being vindicated. He does it by anticipating Oliver's vindication. Here, vindication is manifestly a vicarious attitude: he thinks Oliver will vindicate his trust in him. And Mr. Grimwig also seeks to have his judgment vindicated. But Dickens couches this in a more selfish and therefore somewhat more spiteful language:

> The spirit of contradiction was strong in Mr. Grimwig's breast, at the moment; and it was rendered stronger by his friend's confident smile . . . It is worthy of remark, as illustrating the importance we attach to our own judgments, and the pride with which we put forth our most rash and hasty conclusions, that, although Mr. Grimwig was not by any means a bad-hearted man, and though he would have been unfeignedly sorry to see his respected friend duped and deceived, he really did most earnestly and strongly hope at that moment, that Oliver Twist might not come back. (Dickens 1961: 137)

The scene is exemplary because the reader too looks to see Oliver vindicated and Mr. Grimwig thwarted. But there is more than one person we are anxious about: we want to see Oliver vindicated, especially in the face of Mr. Grimwig's provoking certainty—that'll show him! We want to see Oliver vindicated before Mr. Brownlow too, but mainly perhaps we want to see Mr. Brownlow's trust in Oliver vindicated. We resent Mr. Grimwig's selfish preference for his own judgment to Mr. Brownlow's humane sense of Oliver. (Remember that what we feel is Mr. Grimwig's selfishness is selfish in a psychological sense, in no way a rational or biological one, and indeed derives from what Hume calls a general propensity of benevolence toward mankind.)

Mr. Grimwig's provoking skepticism has consequences: had he not doubted Oliver Mr. Brownlow would not have attempted to vindicate him by ostentatiously entrusting the errand to him, and Nancy and Bill would not have found him on the way to the bookseller. Our anger with them is exacerbated into fury by our frustration that Mr. Grimwig will think himself vindicated, that Mr. Brownlow will be disappointed, and that both will wrongly conclude that Oliver is depraved, Mr. Grimwig triumphantly, Mr. Brownlow sorrowfully. Dickens expertly elicits this frustration in the next scene when he has the bystanders disbelieve Oliver as Nancy and Bill abduct him:

> "What's the matter, ma'am?" inquired one of the women.
>
> "Oh, ma'am," replied the young woman, "he ran away, near a month ago, from his parents, who are hard-working and respectable people; and went and joined a set of thieves and bad characters; and almost broke his mother's heart."
>
> "Young wretch!" said one woman.
>
> "Go home, do, you little brute," said the other.
>
> "I am not," replied Oliver, greatly alarmed. "I don't know her. I haven't any sister, or father and mother either. I'm an orphan; I live at Pentonville."
>
> "Only hear him, how he braves it out!" cried the young woman. (Dickens 1961: 143)

The manifest injustice of the two women stands for the general injustice of those who judge Oliver throughout the book, and that injustice is palpable to readers. And yet what are they but mistaken altruistic punishers, cooperating with what they take to be the imperatives of the urchin's family? And here we can judge how vexatious it is to see altruistic punishment gone wrong: how mistaken or unjust altruistic punishment becomes itself something that we want to see altruistically punished.

It's important to note that it's we who feel this way, not Oliver. He doesn't resent the injustice. We resent it for him. He feels only alarm, and it is a tribute to our sense of his essential innocence that, to use Philip Fisher's phrase, we volunteer the affect for him. Oliver himself is not vindictive, nor does he even think to seek vindication. But we are close to vindictiveness ourselves by this point, and while we will find some satisfaction in the fate of Bill Sikes, it's a good thing that Dickens doesn't cherish a grudge against these bystanders, since they will never learn the truth and repent.

I want to call attention to the following elements of this scene. We are concerned about several interconnected things at once. We worry about Oliver's fate. Part of that worry is about his relation to Mr. Brownlow. We worry therefore about whether Oliver will vindicate Mr. Brownlow's trust. But there has to be an audience for that vindication within the world where it takes place. Our knowledge of Oliver's innocence is not enough. That audience is often primarily the person whose faith we hope to see vindicated (from Odysseus to Gilbert Redman), but there is more pleasure to be anticipated in seeing a vexatious skeptic like Mr. Grimwig get his comeuppance. Mr. Grimwig would be the witness of his own confutation. He would also be the necessary witness—not of Oliver's vindication, but of the vindication of Mr. Brownlow's faith in Oliver.

So we sympathize, or volunteer, or hope to volunteer affect not only for Oliver, and not only for the sympathizing Mr. Brownlow, but—oddly enough—for Mr. Grimwig. We care about what he feels: we want him to see that he has been wrong. We are outraged that he doesn't know it. If we want Mr. Brownlow's faith vindicated, we also anticipate some vindictive pleasure in making Mr. Grimwig eat his head (as he says he will), or his words. Notice that Dickens says of Mr. Grimwig that he is not a bad-hearted man and would be sorry to see his friend duped. Dickens has plenty of bad-hearted men, but Mr. Grimwig has a trickier function: we care about what he feels rather than just wishing him away, as we have wished away the awful magistrate Mr. Fang. Dickens has to balance him delicately between selfish wrongheadedness and a capacity to enter into relations of sympathy. (Scrooge, of course, is another such figure.) When Mr. Grimwig returns later on, now to help punish the tormentors of Oliver, enough time has gone by, a sufficient number of actually selfish characters have appeared, that we feel indulgently disposed toward him and happy to see him absorbed into the general good fellowship of prosocial characters.

Note too that such a capacity for sympathy belongs as well even to the bystanders who excoriate Oliver. Their anger is volunteered on behalf of the mother whose love for Oliver they understand without sharing. Of course they take pleasure in excoriating him, and of course this pleasure is selfish or spiteful. But the alibi for that pleasure is also its real motive: the sense of punishing someone who has done a real wrong (as they think) to his mother.

As a final observation about this sequence, it's worth remarking the narrative's own tender observation of Oliver's ignorance of all this—an ignorance like that of the infant in Adam Smith's (2004) example of the anxious mother, which I quoted in the previous chapter. We sympathize with Oliver, although he is ignorant of the extent of the misfortune we ascribe to him. We resent the wrong done him behind his back, the wrong done him through the misjudgments of Mr. Grimwig and, as we fear, of Mr. Brownlow as well. After retelling Mr. Brownlow's disappointment in Oliver (and Mrs. Bedwin's continued faith in him) upon the testimony of the egregious Mr. Bumble, the narrative insists on Oliver's ignorance of the disappointment he has been the occasion for, and the misconstructions put upon him, in a single-sentence allusion to his feelings: "Oliver's heart sank within him, when he thought of his good kind friends; it was well for him that he could not know what they had heard, or it might have broken outright" (Dickens 1961: 164). His heart would have broken in sympathy for them and for their disappointment, and not in self-pity. Vicarious experience outweighs his own situation.

Likewise, the subjunctive mood of this statement ("it might have") nevertheless elicits our sympathy for Oliver (despite the fact that his heart is not broken) since, in fact, he could not have known what they felt. We sympathize with his capacity to sympathize, with the fact that he might have had his heart broken by the heartbreak he causes, and these reciprocal vicarious feelings, their tendency to volunteer affect for each other, is just what gets us to volunteer affect for them.

I noted above that Mr. Grimwig isn't shown up and shamed for his skepticism about Oliver. There are characters who fulfill that role, especially Mr. and Mrs. Bumble, and Monks, who is forced to confess. Mr. Grimwig instead fills the role of willing acknowledger of his own wrong—the role I saw exemplified above in Caldicott and Charters (in Hitchcock, *The Lady Vanishes* [1998]), and in Laertes. In comedy one might instance Don Pedro, the witness to all things (as Albany is in *King*

Lear; see below), and in another way perhaps Orsino and Olivia (in *Twelfth Night*). Even Edmund may be mentioned here, as opposed to Goneril and Regan; although he might best be compared, in the unexpected schematic I am unfolding, to Nancy, as the repentant wrongdoer who understands his wrong. (Lear and Gloucester share these characteristics as well, but in a way more like Mr. Brownlow.)

Such characters are often called "window characters" in screenwriting (Decker 1998), but window characters may be of considerable affective weight in a story; they're not just confidants (like Horatio or Kent). In Austen's *Persuasion* (1996b) Lady Russell, who has wrongly preferred Elliot to Wentworth, also plays this role, in a passage that Dickens might have been echoing:

> The only one among [Anne's friends], whose opposition of feeling could excite any serious anxiety was Lady Russell. Anne knew that Lady Russell must be suffering some pain in understanding and relinquishing Mr Elliot, and be making some struggles to become truly acquainted with, and do justice to Captain Wentworth. This however was what Lady Russell had now to do. She must learn to feel that she had been mistaken with regard to both; that she had been unfairly influenced by appearances in each; that because Captain Wentworth's manners had not suited her own ideas, she had been too quick in suspecting them to indicate a character of dangerous impetuosity; and that because Mr Elliot's manners had precisely pleased her in their propriety and correctness, their general politeness and suavity, she had been too quick in receiving them as the certain result of the most correct opinions and well-regulated mind. There was nothing less for Lady Russell to do, than to admit that she had been pretty completely wrong, and to take up a new set of opinions and of hopes.
>
> There is a quickness of perception in some, a nicety in the discernment of character, a natural penetration, in short, which no experience in others can equal, and Lady Russell had been less gifted in this part of understanding than her young friend. But she was a very good woman, and if her second object was to be sensible and well-judging, her first was to see Anne happy. She loved Anne better than she loved her own abilities; and when the awkwardness of the beginning was over, found little hardship in attaching herself as a mother to the man who was securing the happiness of her other child. (Austen 1996b: 280–281)

Anne's own discernment and love are vindicated before Lady Russell, who has to admit to herself that she was wrong (as well as before Sir

Walter and Elizabeth when the truth of Mrs. Clay's motives comes out). Anne's anxiety about whether Lady Russell will persist in her opposition to Anne's preference is an anxiety to change how she thinks; she doesn't want to show Lady Russell up, but *does* want her to know that she has been wrong. This is vindication without vindictiveness, and such vindication is always gratifying to a reader, especially if the vindictiveness can find some other target, such as Mr. Elliot and Mrs. Clay, both of whom are more or less banished from the society they seek so assiduously to cultivate. Mrs. Smith, while a relatively minor character, makes all this happen through the gossip she receives and passes on, in conformity with her own desire to punish Mr. Elliot for his treatment of her husband and then of her, a punishment from which she anticipates little gain. But it is Lady Russell who presents the most interesting case here, and the one we feel Austen praises the most. She is purely generous, preferring to see her friend happy to seeing her own judgment vindicated (although that judgment too springs from altruistic propensities), and this suggests the extent to which vindication (and altruistic punishment) are in the service of strong reciprocity, and not ends in themselves.

One interesting aspect of *Oliver Twist* is the virtual disappearance of Oliver from the second half of the book. He has become a Maguffin, the catalyst for a plot where his disparagers, his vindicators and the witnesses of his vindication, become the central characters. We could say much the same about Cordelia in *King Lear;* she is vindicated offstage, and her absence from the plot functions much like Oliver's. So too does Hermione's in *The Winter's Tale;* the oracle vindicates her but the vindication isn't real until Leontes spends the next sixteen years ratifying it. In stories of the vindication of the innocent, from Cordelia to Hermione to Oliver to *Dial M for Murder* (Hitchcock 1987), much of the anxiety that we feel is about whether the altruistic punishers—the police or patrons or fathers—will come to see that they have mistrusted wrongly, have punished wrongly. Will they now make it right? Doing so will, in its turn, vindicate them before those they have wronged. This may be so even if the wronged are dead ("I killed the slave that was ahanging thee," Lear assures the dead Cordelia). In *Hamlet* we may say that the revenger attempts to vindicate the ghost, the nearly absent Maguffin who watches the story of his son's vengeance. But to be absent (especially in Act V) he has to be first present in the plot.

The witnessing ghost is a general feature of revenge tragedy, one

Shakespeare altered to his own fascinating ends by making it part of the drama, not simply introduced in an induction or frame narrative as its most interested audience (like Don Andrea in Kyd's *The Spanish Tragedy* [1999]). We have seen that vindication *in* the world of the fiction matters more than vindication *to* the extra-fictional audience (us), and bringing the ghost closer to membership in that fictional world, whether by an induction scene or by actual absorption into the story, helps with our satisfaction in the vindication.[4] But equally we may say that witnesses even of their own vindication have a centrifugal tendency outward, toward the audience's vicarious experience of their own stories, with the ghost of Talbot that Nashe imagines watching the Talbot scenes, and the audience's reaction to them, in the first part of *Henry VI*.

An eccentric version of the ghostly absence of the figure before whom we might wish to see the hero vindicate him- or herself, may be seen in the death of Gloucester in *King Lear.* He lives to see Edgar in his touch, but not to see whether Edgar will live or die in his conflict with Edmund. The psychological peculiarity of Gloucester's death goes unremarked in the criticism. By psychological peculiarity I mean the question how this death could fit into the psychology of satisfaction in well-formed narrative. The fact that it goes unremarked means that it does fit into such a psychology, and that *King Lear* is indeed a well-formed narrative. But why?

There are of course other reasons for Gloucester's offstage death. He has to die, and in this efficient way he doesn't distract from the climactic deaths of Lear and Cordelia. But I think he is a more central character than this economizing reading would suggest, and we should say that it's appropriate to the tragic ironies he suffers throughout that he should die before witnessing Edgar's triumph and the discomfiture of the evil children. He has predicted that he will live to see their defeat (see below), and this prediction, like every prediction of vindication in *King Lear* ("I will do such things—/ What they are, yet I know not: but they shall be / The terrors of the earth") is falsified (Shakespeare, *King Lear* 2001: II.ii.472–474). We feel, however, that Edgar's vindication, and Gloucester's too, is well-witnessed, most of all perhaps by Albany whose character as witness is central to his presence in the play (even when he witnesses, and makes her witness, his own decisive and exhilarating refutation of Goneril through his production of her letter to Edmund).

Why is it Albany who witnesses Edgar's vindication and therefore Gloucester's as well, and not Gloucester himself? Albany is a character we like, but not one we're particularly concerned about protecting or seeing vindicated. Why should he do the narrative work of witnessing vindication that we might expect Gloucester to do?

This is the lesson about the difficulty of vindication that *Lear* teaches. Vindication, to be at all effective, requires some witness who paradoxically must not be the ideal witness that the injured and now vindicated party would be. The nonideal witness will not raise the otherwise inevitable anxiety of whether vindication is possible, for in the end, the very inextricability of vindication with human cooperation, even with one's enemies, and with the cooperative instincts means that it cannot provide private satisfaction, the private satisfaction that the injured party would feel. The contemplated pleasure of such satisfaction may be the incentive rigged up for us to seek vindication, but vindication will never satisfy the pleasure we anticipate from it. Vindication is concerned with how others feel, those who have done wrong, and seeks to make them know that they are wrong. But by the simplest of paradoxes, if you know you're wrong, you're not wrong any more. If you acknowledge the interests and rights of the person you have wronged, that acknowledgment supports their interests and rights. The desire for vindication both wants and doesn't want that acknowledgment, and so it cannot be met. In Gloucester's case he is both sinner and sinning, and so it is particularly hard for him to balance the complex diagram of contradictory desires he experiences. What does he want of Edmund? What does he want Edgar to want of him? What does he want of Regan, given Edmund's love for her? Can he punish one without the other? No satisfying vindication of Gloucester that he could witness is possible and so instead Albany has to witness it. He stands for a kind of general capacity to judge what is right, rather than a character highly involved in the story he judges. Oddly, like Mr. Grimwig he presents a kind of parable about the limits of vindication, and about how the impulse toward wanting to witness it will have to merge with a more general benevolence toward others to yield narrative satisfaction to an audience.

Winged Vengeance

I cite *King Lear* (Shakespeare 2001) less for this moral lesson, though, than as the most intense representation available of the passionate drama of vindication that I have been claiming is central to narrative. Here I would like to sum up much of the argument of this book by treating as a model for the parallels between vindication and vindictiveness the last scene of Act III, the scene in which Gloucester's eyes are gouged out.

I want to hazard an unlikely description of this horrendous scene, by calling attention to the altruistic element in the punishing (or spiteful) behavior of Edmund, Regan, and Cornwall. I don't wish to defend them but to shed light on psychological motivations whose plausibility we can recognize here. They do not punish him out of their own altruistic purposes, but they do engage in acts motivated by a human and prosocial disposition to engage in altruistic punishment, one that promotes the ends of their own society against those of the French. From their point of view, Lear's irrational behavior and the boisterousness of his knights makes keeping order more difficult. The reason they want to keep order is, of course, to exercise power; but what they intend to do with power is, conversely, to keep order. Their ambition, like all ambition for power, is prosocial, even if its motivations may be described as power-mad, bloodthirsty, self-serving, or greedy. As Hume says, all these elements of a person's character acknowledge the existence of others and solicit acknowledgment from them. This scene makes that fact clear.

An earlier scene, III.5, sets this scene up and I'll start there. Cornwall tells Edmund that he will be revenged on Gloucester for his complicity in the goals of the French, which include saving and restoring Lear. Edmund's response is bracing but pious: he represents himself to Cornwall as a strong figure, nevertheless worried that what he has to do will open him to remonstrance. The mode of this self-representation is that of noting the costs that punishment draws on the punisher's head. Edmund is no altruistic punisher—he is a defector pure and simple, at least at this point in the play—but he knows the type, so this scene acknowledges the existence of the type, and he knows how to play the role. His customary practice is to represent himself as like his interlocutor, and Cornwall's desire for revenge is vindictive and spiteful (and therefore, as we've seen, on a continuum with altruistic punishment). He imitates Cornwall and counterfeits the character of an altruistic punisher, a character we can see the scene beginning to analyze:

Cornwall: I will have my revenge ere I depart his house.
Edmund: How, my lord, I may be censured, that nature thus gives way to loyalty, something fears me to think of.

Cornwall: I now perceive it was not altogether your brother's evil disposition made him seek his death, but a provoking merit, set a-work by a reprovable badness in himself.
Edmund: How malicious is my fortune, that I must repent to be just! . . . O heavens, that this treason were not, or not I the detector! . . . I will persevere in my course of loyalty, though the conflict be sore between that and my blood.

(Shakespeare, *King Lear* 2001: III.v.1–19)

Edmund distinguishes between nature and loyalty and then reiterates the distinction as one between loyalty and blood. The distinction between loyalty and blood is precisely that which the theorists of the evolution of cooperation and, in particular, altruistic punishment explain: the mental incentives of conscience and pleasure that make us act against the rational interests of ourselves and our relations on behalf of a more abstract justice. Edmund makes himself out to be an altruistic punisher, suffering the costs ("I must repent to be just") of making his own away. For my purposes it's not significant that this is a rationally calculated pose; what matters is what he is posing *as,* the recognizable and praiseworthy figure of the altruistic punisher. This is a category that Cornwall recognizes completely. It's not that he recognizes that, as I am claiming, his desire for revenge is continuous with Edmund's putative desire for justice (that vindictiveness is continuous with the justice that would in the end be vindicated). But he does offer an odd bit of sympathy to Edgar, who seems (in Cornwall's judgment of Edmund's false account of his brother) to be not altogether evil. Provoked by the reprovable badness of Gloucester, which would consist in his being a traitor or defector, Edgar too may have the merit of responding to provocation as a provocatively altruistic punisher, not merely as a self-dealer.[5] I note here too that the moral language that Cornwall uses, and that Edmund accepts, indicates that he does not think of himself, but of Gloucester, as the evildoer. From his point of view, Gloucester has defected.

Two scenes later Cornwall gets his revenge, in a scene that miniaturizes many of the issues of anger, revenge, and punishment, and the observation of those things, that are central to the play and also to my argument. Let's start with anger. When one of Cornwall's servants tries to prevent the mutilation he warns Cornwall not to count on the servant's calculation of his own advantage:

First Servant: Hold your hand, my lord:
I have served you ever since I was a child;
But better service have I never done you
Than now to bid you hold.
Regan: How now, you dog!
First Servant: If you did wear a beard upon your chin,
I'd shake it on this quarrel. What do you mean?
Cornwall: My villain!
[*They draw*]
First Servant: Nay, then, come on, and take the chance of anger.
(Shakespeare, *King Lear* 2001: III.vii.71–78)

The servant acts as an altruistic punisher, attempting to benefit both Gloucester and Cornwall by his intervention. The scene recapitulates the opening of the play: a subaltern stands up to a leader about to act violently, and by coming between the dragon and his wrath draws that wrath on himself. Kent sacrifices himself for Lear and for Cordelia, whose interests Lear fails to recognize as his own; likewise, the servant sacrifices himself for Cornwall's interests, which are also Gloucester's. His altruism is meant to prevent a devolution into the Hobbesean nightmare of pure and violent rivalry, and to preserve the laws of hospitality ("Good my friends, consider / You are my guests: do me no foul play, friends" [Shakespeare, *King Lear* 2001:III.vii.30–31]) that are the most basic elements of human cooperative altruism. The servant warns both Regan and Cornwall of his anger, that is, that he is not calculating his own advantage. As we've seen, anger is the emotional reflex that evolution provides to commit us to actions that go counter to our own individual and immediate interests. The servant's strictly irrational anger means that Cornwall has to take his warning seriously, that he won't be stopped by a reminder of his status: "My villain." Anger serves the prosocial structure of altruistic punishment. This is an attitude and account of anger that Cornwall has already learned from Kent, when he explains his abuse of Oswald:

Cornwall: Peace, sirrah!
You beastly knave, know you no reverence?
Kent: Yes, sir; but anger hath a privilege.
(Shakespeare, *King Lear* 2001: II.ii.69–71)

Cornwall accepts this, as well he might. Although his analysis of Kent's bluntness is shrewd, it isn't reductive. He too has reasonable self-control, but we've already seen (as I've just noted) the contrast between him and the completely calculating Edmund, and it's clear from the events leading to the moment that the servant draws that Cornwall understands anger from the inside, for Cornwall has recognized, rationally enough, that he cannot simply kill Gloucester (although later Regan will regret the costs they pay for not killing him). But Cornwall also recognizes or recalls that anger hath a privilege:

> Though well we may not pass upon his life
> Without the form of justice, yet our power
> Shall do a courtesy to our wrath, which men
> May blame, but not control.
>
> (Shakespeare, *King Lear* 2001: III.vii.24–27)

Cornwall's wrath will certainly become real at the end of this scene, but is it real yet? He's as cold and calculating as ever, and so his passion seems calculated as well. And yet that calculation serves not some utilitarian end but the gratification of a passion. He wants to hurt Gloucester. His wrath may be an excuse he can cite, but it excuses an action or attitude not very different from the wrath itself: a preeminent desire to cause pain to an antagonist or adversary. You can see that this desire is irrational (even if the extent that he will indulge it is nicely calculated) if you compare Cornwall to the far more rational Bullingbrook or Octavius Caesar. Both accurately measure and counter their rivals' responses to their calculated appropriations of power, and both are coolly indifferent to any personal vindication or vindictiveness. Most of Shakespeare's great villains, from Richard III to Shylock to Leontes, want to cause pain and grief, even at the cost of being trapped by their own engines. Iago provides the most vivid illustration of this impulse.[6]

When Regan and Cornwall have him bound, Gloucester repeats his complaint (made earlier to Edmund) that they are violating the laws of hospitality, but this is part of their point. Those laws record a primordial and elemental disposition to cooperate (even in the archeological record from late Paleolithic times). By a strange and natural antithesis their sacred character makes it possible for someone to express primordial and elemental vindictiveness by violating them, as when Laertes declares himself ready to cut Hamlet's throat in the church. The same irreducible and vicarious concern for the other that establishes the laws of hospitality can undermine them; to put it another way, it is to act in a paradoxical conformity with their sacred character to violate them intentionally, and their sacred character is just that they take the experience and perspective of the other as sacred. I doubt that Regan and Cornwall take the laws as seriously as the servants do ("Bind him, I say," Cornwall has to insist) or as the audience does. But they take them seriously enough to know that Gloucester does, and that part of their revenge is their violation of the compact.

As for their own passion, you can see how Regan's and Cornwall's angry desire to hurt Gloucester overrides calculation by the way the scene unfolds. Cornwall wants to know where Gloucester has sent Lear:

> *Gloucester:* To Dover.
> *Regan:* Wherefore to Dover? Wast thou not charged at peril—
> *Cornwall:* Wherefore to Dover? Let him answer that.
>
> (Shakespeare, *King Lear* 2001: III.vii.50–52)

Regan asks the question "Wherefore to Dover?" as a kind of rhetorical flourish: How dare you? She does not stay for an answer, seeking instead to rebuke him. We don't notice this until Cornwall interrupts her by repeating the question—"Let him answer that"—reminding us that there's important information he wishes to get: who are the well-armed friends Lear and Gloucester have there, how many, with what relation to the already landed French forces, what lines of communication have been set up among them, and so on.[7] Cornwall more than Regan is interrogating Gloucester for some rational purpose, a purpose she joins by posing the question more respectfully a moment later: "Wherefore to Dover, sir?"[8] But that purpose quickly gets lost again in the passions of the moment.

> *Regan:* Wherefore to Dover, sir?
> *Gloucester:* Because I would not see thy cruel nails
> Pluck out his poor old eyes; nor thy fierce sister
> In his anointed flesh stick boarish fangs.
> The sea, with such a storm as his bare head
> In hell-black night endured, would have buoy'd up,
> And quench'd the stelled fires:
> Yet, poor old heart, he holp the heavens to rage.
> If wolves had at thy gate howl'd that dern time,

> Thou shouldst have said "Good porter, turn the key,"
> All cruels else subscribe: but I shall see
> The winged vengeance overtake such children.
>
> (Shakespeare, *King Lear* 2001: III.vii.54–65)

Gloucester's answer is intense but not informative. Whether adroitly or accidentally, he avoids divulging the strategic reason he sent Lear to Dover and substitutes a highly charged figurative one instead. Note that his reply answers the question why he sent Lear away—to save him from Regan—but not why he sent Lear to Dover in particular. That we tend not to notice how he evades the question testifies to the success with which he evades it.

And in fact he has so enraged Cornwall that even Cornwall forgets to pursue the question, as his wrath certainly becomes genuine and complete:

> *Cornwall:* See't shalt thou never.
>
> (Shakespeare, *King Lear* 2001: III.vii.66)

The situation has devolved into a contest not between rational calculators but between promoters of vengeance. Gloucester divulges what he would not see—further harm to Lear—and what he trusts he shall see—vengeance on Cornwall, Regan, and Goneril. This is manifestly an altruistic desire: he defies his enemies by contemplating not his own vindication but Lear's as the goddess of vengeance (Dike's double) overtakes such children. Gloucester will live to observe the vindication take place; Cordelia and Lear will be vindicated in the world in which they have been wronged and their vindication demonstrated to those there to know it, and Gloucester expects that he will be such a person.

In *saying* so to Cornwall and Regan he grafts his own vindication onto that of Lear and Cordelia. When they are vindicated so too will his anticipation of their vindication be vindicated in its turn. And we become part of this cooperating or at least well-wishing (strongly reciprocating) chain of vindicated observers, as we expect to see Gloucester's prophecy vindicated. Or to telescope these different orders back together, we can simply say that Gloucester's defiance here presents a central and absolutely plausible example of the altruistic wish to see someone punished, and acts as a proxy for the audience's identical desire. Our own investment in the plot of the play is by this point an investment in seeing such children get what's coming to them.

We can understand now the almost routine allusions to blinding in the play as part of a general and shared expectation that the ability to monitor what is happening to others is part of the fabric of punishment, which requires: 1) the ability to monitor punishment (you get punished for taking the side of one who is out of favor, who is the one being punished), including most importantly for literature, 2) the ability to monitor the punishment of those who have themselves punished unjustly or vindictively ("Who stock'd my servant? Regan, I have good hope / Thou didst not know on't"), capped by 3) the vindication of the vindictively punished (Shakespeare, *King Lear* 2001: II.iv. 180–181). Everyone takes everyone else's ability to see seriously because of what they can monitor by seeing and not only because of the exceeding and dramatically effective horror of contemplating the loss of one's eyes. That goes without saying, or certainly without so much saying. What doesn't go without saying, what the play takes as one of its themes, is the priority, over claims rationally more pressing, of witnessing the vindication of the wronged. The strange and powerful fact that Gloucester *doesn't,* in fact, live to see the winged vengeance overtake such children should be arrayed among the significant illusions of human meaning or hopefulness that the play dissipates. (This also is part of the punishment that Gloucester deserves, in Edgar's unsparing judgment later on, and as we've seen it solves a problem in the gratifications of narrative as well.)

Gloucester does get to see a small part of the vindication he anticipates though, as the last words of the altruistic servant who gives up his life for both Gloucester and the "better service" of Cornwall attest:

> *First Servant:* O, I am slain! My lord, you have one eye left
> To see some mischief on him. O!
> *Cornwall:* Lest it see more, prevent it. Out, vile jelly!
> Where is thy lustre now?
>
> (Shakespeare, *King Lear* 2001: III.vii.80–83)

The Servant dies unvindicated by the gods ("He that would think to live till he be old, give me some help"), but is able to vindicate Gloucester by *showing* him Cornwall's wound. Cornwall responds in kind, and—spitefully? vindictively? altruistically?—rates preventing Gloucester from seeing him bleed, from seeing how wounded he is, higher than binding the wound immediately. Although there is no reason to suppose that his placing the punishment of Gloucester above tending to his own wounds leads to his death, doing so certainly goes counter to his rational interests. But the lesson he and Regan wish to teach Gloucester is the lesson that all altruistic punishers wish to teach, since their altruism extends not only to society at large but even to the defector as a part of that society (even the punisher cooperates with the traitor):

> *Cornwall:* Lest it see more, prevent it. Out, vile jelly!
> Where is thy lustre now?
> *Gloucester:* All dark and comfortless. Where's my son Edmund?
> Edmund, enkindle all the sparks of nature,
> To quit this horrid act.
> *Regan:* Out, treacherous villain!
> Thou call'st on him that hates thee: it was he
> That made the overture of thy treasons to us;
> Who is too good to pity thee.
> *Gloucester:* O my follies! then Edgar was abused.
> Kind gods, forgive me that, and prosper him!
> *Regan:* Go thrust him out at gates, and let him smell
> His way to Dover.
>
> (Shakespeare, *King Lear* 2001: III.vii.82–93)

If Gloucester's passionate wish is gratifying and unsurprising, perhaps it's no surprise either that these bitterly traded accusations motivate Regan and Cornwall, albeit on the wrong side. If the desire for vengeance and vindication is altruistic when felt by the side of those who are wronged, it is also altruistic when experienced on the part of those who believe themselves wronged, irrespective of whether we think that belief is legitimate. For this reason we can see the passion animating both Cornwall and Regan as transcending self-regard. Regan means seriously the moral judgment she makes when she praises Edmund for betraying Gloucester, against his own "natural" (kinship) interests, as showing that he "is too good to pity thee."

The winged vengeance does overtake these children in the end, and it may be that we could start answering the old and difficult question of why tragedy gives pleasure by saying that all narratives of vindication give pleasure, and that narrative is only narrative if it allows us to anticipate vindication.[9] The bleakness of *Lear* and the essential happiness of *Oliver Twist* gratify our desires in ways essentially alike. (This is why it doesn't matter that Oliver disappears from the book.) If we are not to see Cordelia (or Nancy) live we are permitted to see their persecutors punished, and permitted (to the extent such things are not self-contradictory) to empathize with the surprise some of them feel to find themselves in hell.[10]

Coda

> If the red slayer thinks he slays,
> Or if the slain thinks he is slain,
> They know not well the subtle ways
> I keep, and pass, and turn again.
>
> —Ralph Waldo Emerson, "Brahma"

I have been treating narrative here in two ways—as a verisimilitudinous record of human cooperation and also (since the fact of verisimilitude is good evidence of accuracy because we are so good at monitoring cooperation, of judging what looks true) as an object of the kinds of interest that human cooperation requires and rewards. These two things are related because cooperation is regulated by strongly reciprocating and cooperative monitors.

I *haven't* been discussing much the esthetics of narrative, except somewhat tangentially. But I think at least some portion of those esthetics is relevant here. There is first of all the well-considered argument that esthetics constitutes a kind of costly signal or honest handicap. Sprezzatura, as a central esthetic term, is an old example of this argument. But I want to suggest something simpler as well: that we like to see cooperation in action. The idea of the artist or writer or narrator as gift-giver is a version of such a liking, perhaps (see Hyde 1983). But I think also we like to see actors or dancers or singers or other repertory groups put a story together for us. They cooperate with one another to do so, and we like cooperation. We like to see straight men setting up jokes for their wiseacre partners. We like to see dancers lifting each other up. We like to see cooperation among the actual humans doing the performance. I think here of some of Jean Renoir's movies, like the great *Crime de Monsieur Lange* (2003), and the way we feel that everyone is friends there. In narrative this happens in the great party novels like *Mrs. Dalloway* (Woolf 1925) or Thomas Bernhard's *Woodcutters* (1987). This latter consists in a tirade delivered by a trademark Bernhard crank against the other guests at a Viennese party he is attending, against their loathsome and vapid stupidity, and that of the city itself. After two hundred pages of this, he nevertheless is struck by how much he loves them, and all of Vienna, and we are made strangely happy. (*Woodcutters* is as much about the gratification afforded by narratives rewarding those who keep the laws of hospitality as 3,000 years earlier Homer and Genesis were.) Such an ending reminds me of the therapeutic song, which so struck Claude Lévi-Strauss, sung by the *nele* or shaman among the Cuna Indians of Central America, aiding the mother in difficult labor against the ambiguous deity Muu, whose domain is the mother's uterus, and who forms the fetus but may also seek to capture the soul of the mother. When the *nele* finally defeats her, the song he sings narrating his victory ends with her parting words that "almost sound like an invitation: 'Friend *nele,* when do you think to visit me again?'" (Lévi-Strauss 1976: 187). The end of Mikhail Bulgakov's *Master and Margarita* (1996) is very similar: Yeshua (Jesus) and Woland (Satan) turn out to be friends, to help each other and to find meaning in doing so. (*Paradise Regained* [Milton 1998] toys with a similar friendship.)[1]

These cooperative culminations are versions of the principle of *convergence* that is an explicit part of the structure of the Hollywood screenplay (see Decker 1998), in which all the major characters should ideally return on the last page, a principle largely followed by Shakespeare as well, in *Hamlet* or *Lear* no less than in *The Winter's Tale* or *The Tempest* or any of the early comedies. There everyone is. Even when Oliver Twist or Cordelia are not there, in the end, their *absence* is why everyone else comes.

This pleasure in cooperation is what makes characters like Grimwig so important: they too belong to the dance of cooperation (and it's what makes Dickens so great, since every character seems to consent, cheerily, to be a Dickens character). I would like to think that this book helps communicate the pleasure that I take in reading Grafen (1990a) and the Zahavis (1997) on cooperation even between life-and-death rivals.

In this sense H. P. Grice's cooperative principle (whereby we always try to make sure we're talking about the same things as our interlocutors) would also be relevant, in a way not explicitly contemplated by him: the speakers in a narrative cooperate to make it perspicuous to an audience, and not only to advance the direction or purpose of their own conversation (Grice 1975: 45–46, 49–50). I don't think it's stretching the kind of cooperation I am thinking about to regard such works as Franz Kafka's, from the parables to *Das Schloss*, as partly affording the pleasure of such cooperation, of watching the figurants—the man who seeks the law, the guard who keeps watch faithfully at his gate all those years and then tells him the truth—act together to create the parable itself, as though they volunteer for the purposes of the exposition.

Perhaps the best simple example of this may be found in ballads and balladlike poems, such as Sir Walter Scott's "Proud Masie" (1960). Masie continues to question the robin even after he has told her that she will be wedded to death. Her second question is a repetition of the first. There is no narrative development: it does not come out of disbelief or misunderstanding or desire for clarification. Their interaction is nearly ritualistic, and the questions she asks are questions that make possible the robin's eerie and wonderful answers. They cooperate in this, Masie and the robin, and so we feel her less as a rebuked or punished figure (though she is that too) than as part of the spooky dramatis personae of the poem, a person whose cooperation with the story is not the least element of her spookiness, and of our sense that in fact she will be "welcome," that her *welcome* is finally a more important word than the pride the poem officially undercuts. (That this is a mad song from *The Heart of Midlothian* [Scott 1960] sung by the dying Madge Wildfire, who is certainly not talking about herself and yet is feeling in Masie's death a representation of her own, contributes to the effect.)

At any rate, I want to report this fact: that the pleasure I take from reading the Zahavis (1997) is like, strangely enough perhaps, the pleasure I take from reading Maurice Blanchot (1983). At the end of his early story "L'idylle," the protagonist who has gone through horrendous frustration and disappointment to no end finally gets set to leave the stage; another character, to whom he complains of what has happened, tells him, "That doesn't prevent its having been an idyll." And it doesn't. It was an idyll, in much the same way that there is something idyllic in contemplating the dance of cooperation. Not that prey like the predators they cooperate with, nor perhaps do most species like cooperation at all. They are forced into it. But we like it, and like reading about it and seeing it enacted.

References

Notes

Index

References

Aeschylus. *The Eumenides*. 1960. In *Greek Tragedies*, ed. David Grene et al. Chicago: University of Chicago Press.

Ainslie, George. 1975. "Specious Reward: A Behavioral Theory of Impulsiveness and Control." *Psychological Bulletin* 82: 463–496.

———, 2001. *Breakdown of Will*. New York: Cambridge University Press.

Ainslie, George, and Richard Herrnstein. 1981. "Preference Reversal and Delayed Reinforcement." *Animal Learning and Behavior* 9: 476–482.

Anscombe, G. E. M. 2000. *Intention*. Cambridge, Mass.: Harvard University Press.

Aristophanes. 1943. *The Frogs*. In *Fifteen Greek Plays*. New York: Oxford University Press.

Auerbach, Eric. 1953. *Mimesis: The Representation of Reality in Western Literature*. Princeton, N.J.: Princeton University Press.

Austen, Jane. 1990. *Pride and Prejudice*, ed. James Kinsley. New York: Oxford University Press.

———. 1996a. *Northanger Abbey*. New York: Penguin.

———. 1996b. *Persuasion*. New York: Signet.

———. 1998. *Sense and Sensibility*, ed. James Kinsley. New York: Oxford University Press.

———. 2003a. *Emma*, ed. James Kinsley and Adela Pinch. New York: Oxford University Press.

———. 2003b. *Mansfield Park*, ed. James Kinsley. New York: Oxford University Press.

Austin, J. L. 1970. *Philosophical Papers*. Oxford: Clarendon.

Axelrod, R. 1997. *The Complexity of Cooperation: Agent-Based Models of Competition and Collaboration*. Princeton, N.J.: Princeton University Press.

———. 1984. *The Evolution of Cooperation*. New York: Basic Books.

Axelrod, R., and W. D. Hamilton. 1981. "The Evolution of Cooperation." *Science* 211: 1390–1396.

Baldwin, William. 1960. *The Mirror for Magistrates*, ed. Lily B. Campbell. New York: Barnes and Noble.

Barash, David P., and Nannelle R. Barash. 2005. *Madam Bovary's Ovaries: A Darwinian Look at Literature.* New York: Delacorte.

Basu, Kaushik, 2007. "The Traveler's Dilemma," *Scientific American* (June): 90–95.

Bataille, Georges. 1991. *The Accursed Share: An Essay on General Economy.* New York: Zone Books.

Beckett, Samuel. 1956. *Malone Dies.* New York: Grove Press.

Benjamin, Walter. 1979. "On the Mimetic Faculty." In *One-way Street, and Other Writings.* London: NLB.

Bernhard, Thomas. 1987. *Woodcutters.* New York: Knopf.

Bersani, Leo, and Ulysse Dutoit. 1985. *The Forms of Violence: Narrative in Assyrian Art and Modern Culture.* New York: Schocken.

Bickerton, Dereck. 2000. "Reciprocal Altruism as the Predecessor of Argument Structure." In *Lingua ex machina: Reconciling Darwin and Chomsky with the Human Brain,* ed. William H. Calvin and Dereck Bickerton, pp. 123–134. Cambridge, Mass.: MIT Press.

Blanchot, Maurice. 1983. "L'idylle." In *Après coup; précédé par Le ressassement éternal.* Paris: Minuit.

Bliege Bird, Rebecca, Eric Alden Smith, and Douglas W. Bird. 2001. "The Hunting Handicap: Costly Signaling in Human Foraging Strategies." In *Behavioral Ecology and Sociobiology,* 50 (1): 9–19.

Bloom, Harold. 1973. *The Anxiety of Influence: A Theory of Poetry.* New York: Oxford University Press.

———. 1998. *Shakespeare: The Invention of the Human.* New York: Riverhead.

Booker, Christopher. 2004. *The Seven Basic Plots: Why We Tell Stories.* New York: Continuum.

Boone, James L. 1998. "The Evolution of Magnanimity: When Is It Better to Give Than to Receive." *Human Nature* 9(1): 1–21.

Borges, Jorge Luis. 1969. "Three Versions of Judas." In *Ficciones,* ed. and with an introduction by Anthony Kerrigan, pp. 151–158. New York: Grove.

Bourdieu, Pierre. 1984. *Distinction: A Social Critique of the Judgement of Taste.* Cambridge, Mass.: Harvard University Press.

Boswell, James. 1998. *The Life of Samuel Johnson, LL.D.* New York: Oxford University Press.

Bowles, Samuel, Jung-Kyoo Choic, and Astrid Hopfensitzd. 2003. "The Co-Evolution of Individual Behaviors and Social Institutions." *Journal of Theoretical Biology* 223: 135–147.

Boyd, Brian. 2001. "The Origin of Stories: *Horton Hears a Who.*" *Philosophy and Literature* 25: 197-214.

———. 2005. "Evolutionary Theories of Art." In *The Literary Animal: Evolution and the Nature of Narrative,* ed. Jonathan Gottschall and David Sloan Wilson, pp. 147–176. Evanston, Ill.: Northwestern University Press.

———. 2006. "Getting It All Wrong." *American Scholar* 75(4): 18–30.

Boyd, Robert, Herbert Gintis, Samuel Bowles, and Peter J. Richerson. 2003. "The Evolution of Altruistic Punishment." *PNAS* 100, no. 6 (March 18): 3531–3535.

Boyd, Robert, and Peter J. Richerson. 2005. *The Origin and Evolution of Cultures*. New York: Oxford University Press.

Boyer, Pascal. 2001. *Religion Explained: The Evolutionary Origins of Religious Thought*. New York: Basic Books.

Brown, Donald E. 1991. *Human Universals*. New York: McGraw-Hill

Browning, Robert. 1911. *The Ring and the Book*. New York: Dutton.

Bruner, Jerome. 1990. *Acts of Meaning*. Cambridge, Mass.: Harvard University Press.

Buchler, Ira R., and Hugo G. Nutini, eds. 1969. *Game Theory in the Behavioral Sciences*. Pittsburgh: University of Pittsburgh Press.

Bulgakov, Mikhail. 1996. *The Master and Margarita*, trans., Richard Pevear and Larissa Volokhonsky. New York: Vintage.

Buller, David J. 2005. *Adapting Minds: Evolutionary Psychology and the Persistent Quest for Human Nature*. Cambridge, Mass.: MIT Press.

Caillois, Roger. 1987. *Le mythe et l'homme*. Paris: Flammarion

Calvin, William H., and Dereck Bickerton. 2000. *Lingua ex machina: Reconciling Darwin and Chomsky with the Human Brain*. Cambridge, Mass.: MIT Press.

Camerer, Colin. 2003. *Behavioral Game Theory: Experiments in Strategic Interaction*. Princeton, N.J.: Princeton University Press.

Campbell, Donald P. 1987. "Evolutionary Epistemology." In *Evolutionary Epistemology, Rationality, and the Sociology of Knowledge*, ed. Gerald Radnitzky and W. W. Bartley III, pp. 47–89. La Salle, Ill.: Open Court.

Carroll, Joseph. 2004. *Literary Darwinism: Evolution, Human Nature, and Literature*. New York: Routledge.

Carroll, Noël. 1990. *The Philosophy of Horror*. New York: Routledge.

———. 1998. *A Philosophy of Mass Art*. New York: Oxford University Press.

Cavell, Stanley. 2002. *Must We Mean What We Say?* Updated ed. New York: Cambridge University Press.

———. 2003. *Disowning Knowledge in Seven Plays of Shakespeare*, pp. 39–124. New York: Cambridge University Press.

Cervantes, Miguel de. 2003. *Don Quixote*, trans. Edith Grossman, New York: Ecco.

Chaucer, Geoffrey. 1989. *The Canterbury Tales: Fifteen Tales and the General Prologue: Authoritative Text, Sources and Backgrounds, Criticism*, ed. V. A. Kolve and Glending Olson. New York: W. W. Norton.

Christie, Agatha. 1983. *The Murder of Roger Akroyd*. New York: Bantam Books.

———. 2002. *Witness for the Prosecution*. New York: HarperCollins.

Coetzee, J. M. 2004. "What Philip Knew." *New York Review of Books* 51, no. 18 (November 18): 4–6.

Coleridge, Samuel Taylor. 1969. *Biographia Literaria.* In *The Collected Works of Samuel Taylor Coleridge,* ed. L. Patton and P. Mann. Princeton, N.J.: Princeton University Press.

Conrad, Joseph. 1998. *Heart of Darkness and Other Tales,* ed. Cedric Watts. New York: Oxford University Press.

Crews, Frederick. 1999. *Unauthorized Freud: Doubters Confront a Legend.* New York: Penguin.

Cronin, Helena, and John Maynard Smith. 1993. *The Ant and the Peacock: Altruism and Sexual Selection from Darwin to Today.* New York: Cambridge University Press.

Damisch, Hubert. 1994. *The Origin of Perspective,* trans. John Goodman. Cambridge, Mass.: MIT Press.

Dante Alighieri. 2003. *Purgatorio.* New York: Doubleday.

Darwin, Charles. 1998. *Expression of the Emotions in Man and Animals.* London: HarperCollins.

———. 2004. *The Descent of Man, and Selection in Relation to Sex.* New York: Penguin.

Dawkins, Richard. 2006. *The Selfish Gene.* 3rd ed. New York: Oxford University Press.

Dawkins, R., and J. R. Krebs. 1978. "Animal Signals: Information or Manipulation?" In *Behavioral Ecology,* ed. J. R. Krebs and N. B. Davies, pp. 282–309. Oxford: Blackwell.

De Man, Paul. 1984. *The Rhetoric of Romanticism.* New York: Columbia University Press.

Decker, Dan. 1998. *Anatomy of a Screenplay: Writing the American Screenplay from Character Structure to Convergence.* Chicago: Screenwriters Group.

Deleuze, Gilles. 1977. "Nomad Thought." In *The New Nietzsche,* ed David Allison, pp. 142–149. New York: Dell.

Denant-Boèmont, Laurent, David Masclet, and Charles Noussair. 2005. "Punishment, Counterpunishment and Sanction Enforcement in a Social Dilemma Experiment." Available at http:halshs.ccsd.cnrs.fr/docs/00/06/29/77/PD publicgood_punishment40.pdf.

Dennett, Daniel. 1995. *Darwin's Dangerous Idea: Evolution and the Meanings of Life.* New York: Simon & Schuster.

Dickens, Charles. 1961. *Oliver Twist.* New York: Signet.

———. 1966. *Hard Times: An Authoritative Text, Backgrounds, Sources, and Contemporary Reactions, Criticism.* New York: W. W. Norton.

———. 1977. *Bleak House: An Authoritative and Annotated Text, Illustrations, a*

Note on the Text, Genesis and Composition, Backgrounds, Criticism, ed. George Harry Ford and Sylvere Monod. New York: W. W. Norton.

———. 1996. *A Christmas Carol.* In *Christmas Books.* New York: Oxford University Press.

Dowd, Maureen. 1996. *Liberties. New York Times,* December 1, "Week in Review," p. 9.

Doyle, Arthur Conan, Sir. 1902. *A Study in Scarlet* and *The Sign of the Four.* New York: D. Appleton.

Eatwell, John, Murry Milgate, and Peter Newman, eds. 1989. *The New Palgrave: Game Theory.* New York: W. W. Norton.

Edelman, Gerald M. 1987. *Neural Darwinism: The Theory of Neuronal Group Selection.* New York: Basic Books.

Eliot, George. 1977. *Middlemarch: An Authoritative Text, Backgrounds, Reviews, and Criticism.* New York: W. W. Norton.

———. 1999. *Daniel Deronda,* ed. Graham Handley. New York: Oxford University Press.

Empson, William. 1951. *The Structure of Complex Words.* New York: New Directions.

Fanfiction Glossary. Available at http:www.subreality.com/glossary/terms.htm.

Faulkner, William. 1950. *Collected Stories.* New York: Random House.

Fehr, Ernst. 2002. "The Economics of Impatience." *Nature* 415: 269–272.

Fehr, Ernst, and U. Fischbacher. 2003. "The Nature of Human Altruism—Proximate and Evolutionary Origins." *Nature* 425: 785–791.

———. 2004. "Third-Party Punishment and Social Norms." *Evolution and Human Behavior* 25: 63–87.

Fehr, Ernst, and Simon Gächter. 1998. "Reciprocity and Economics: The Economic Implications of Homo Reciprocans." *European Economic Review* 42: 845–859.

———. 2000. "Cooperation and Punishment in Public Goods Experiments." *American Economic Review* 90, no. 4 (September): 980–994.

———. 2002. "Altruistic Punishment in Humans." *Nature* 415: 137–140.

Fehr, Ernst, and J. Henrich. 2003. "Is Strong Reciprocity a Maladaptation? On the Evolutionary Foundations of Human Altruism." In *The Genetic and Cultural Evolution of Cooperation,* ed. P. Hammerstein, pp. 55–82. Cambridge, Mass.: MIT Press.

Fisher, Philip. 2002. *The Vehement Passions.* Princeton, N.J.: Princeton University Press.

Fisher, R. A. 1915. "The Evolution of Sexual Preferences." *Eugenics Review* 7: 184–192.

———. 1958. *The Genetical Theory of Natural Selection.* New York: Dover.

Flack, Jessica C., Frans B. M. de Waal, and David C. Krakauer. 2005. "Social Structure, Robustness, and Policing Cost in a Cognitively Sophisticated Species." *American Naturalist* 165(5): E126–E139.

Flesch, William. 1992. *Generosity and the Limits of Authority: Shakespeare, Herbert, Milton.* Ithaca, N.Y.: Cornell University Press.

———. 2001. "The Conjuror's Trick, or, How to Rhyme." *The Literary Imagination* 3(2): 184–204.

Fodor, Jerry. 2001. *The Mind Doesn't Work That Way: The Scope and Limits of Computational Psychology.* Cambridge, Mass.: MIT Press.

———. 2005. *The Selfish Gene Pool* (review of Buller). *TLS* 5339 (July 29).

Foucault, Michel. 1977. *Discipline and Punish: The Birth of the Prison.* New York: Pantheon.

Frank, Robert H. 1988. *Passions within Reason: The Strategic Role of the Emotions.* New York: W. W. Norton.

Frank, S. A. 1995. "Mutual Policing and Repression of Competition in the Evolution of Cooperative Groups." *Nature* 377: 520–522.

Frankfurt, Harry. 2004. *The Reasons of Love.* Princeton, N.J.: Princeton University Press.

Freud, Sigmund. 1905. "Psychopathic Characters on the Stage." Vol. 7, *The Standard Edition of the Complete Psychological Works of Sigmund Freud,* trans. James Strachey et al., pp. 305–310. Repr. 1957; New York: W. W. Norton.

———. 1907. *Delusion and Dream in Jensen's "Gradiva."* Vol. 9, *The Standard Edition of the Complete Psychological Works of Sigmund Freud,* trans. James Strachey et al., pp. 7–95. Repr. 1957; New York: W. W. Norton.

———. 1909. "Family Romances." Vol. 9, *The Standard Edition of the Complete Psychological Works of Sigmund Freud,* trans. James Strachey et al., pp. 235–244. Repr. 1957; New York: W. W. Norton.

———. 1913. *Totem and Taboo.* vol. 13, *The Standard Edition of the Complete Psychological Works of Sigmund Freud,* trans. James Strachey et al., pp. 1–161. Repr. 1957; New York: W. W. Norton.

———. 1921. *Group Psychology and the Analysis of the Ego.* vol. 18, *The Standard Edition of the Complete Psychological Works of Sigmund Freud,* trans. James Strachey et al., pp. 69–144. Repr. 1957; New York: W. W. Norton.

Frye, Northrop. 1957. *Anatomy of Criticism: Four Essays.* Princeton, NJ: Princeton University Press.

Gallese, Vittorio and Alvin Goldman. 1998. "Mirror neurons and the simulation theory of mind-reading." *Trends in Cognitive Sciences* 2: 493-501.

Gintis, H. 2000. "Strong Reciprocity and Human Sociality." *Journal of Theoretical Biology* 206: 169–179.

Gintis, H., S. Bowles, R. Boyd, and E. Fehr. 2003. "Explaining Altruistic Behavior in Humans." *Evolution and Human Behavior* 24: 153–172.

Gintis, H., E. Smith, and S. Bowles. 2001. "Costly Signaling and Cooperation." *Journal of Theoretical Biology* 213: 103–119.

Goffman, Erving. 1981. *Forms of Talk*. Philadelphia: University of Pennsylvania Press.

Gopnik, Alison, A. N. Meltzoff, and P. K. Kuhl. 1999. *The Scientist in the Crib: Minds, Brains and How Children Learn*. New York: Harper Collins.

Gopnik, Alison. 2007. "Cells That Read Minds?" *Slate,* April 26. Available at www.slate.com/id/2165123/

———. 1993. "How We Know Our Minds: The Illusion of First-Person Knowledge of Intentionality." *Behavioral and Brain Sciences* 16: 1–14.

Gottschall, Jonathan, and David Sloan Wilson, eds. 2005. *The Literary Animal: Evolution and the Nature of Narrative*. Evanston, Ill.: Northwestern University Press.

Gould, Stephen Jay. 1977. *Ontogeny and Philogeny*. Cambridge, Mass.: Harvard University Press.

———. 2002. *The Structure of Evolutionary Theory*. Cambridge, Mass.: Harvard University Press.

Gould, S. J., and Richard Lewontin. 1979. "The Spandrels of San Marco and the Panglossian Paradigm: A Critique of the Adaptationist Programme." *Soc Proceedings of the Royal Society of London, Series B* 205(1161): 581–598.

Grafen, Alan. 1990a. "Biological Signals as Handicaps." *Journal of Theoretical Biology* 144: 517–546.

———. 1990b. "Sexual Selection Unhandicapped by the Fisher Process." *Journal of Theoretical Biology* 144: 473–516.

Gray, John. 2007. "Are We Born Moral?" *New York Review of Books*, 54, no. 8 (May 10): 26–28.

Greene, Graham. 1938. *Brighton Rock*. New York: Viking.

Grice, H. P. 1975. "Logic and Conversation." In *Speech Acts,* ed. Peter Cole and Jerry L. Morgan, pp. 41–58. New York: Academic Press.

Grosz, Elizabeth. 2005. *Nick of Time: Politics, Evolution, and the Untimely*. Durham, N.C.: Duke University Press.

Guilford, T., and M. S. Dawkins. 1991. "Receiver Psychology and the Evolution of Animal Signals." *Animal Behaviour* 42: 1–14.

Gürerk, Özgür, Bernd Irlenbusch, and Bettina Rockenbach. 2006. "The Competitive Advantage of Sanctioning Institutions." *Science* 312 (April 7): 108–111.

Güth, Werne, and Eric Van Damme. 1998. "Information, Strategic Behavior, and Fairness in Ultimatum Bargaining: An Experimental Study." *Journal of Mathematical Psychology* 42: 227–247.

Hamilton, W. D. 1963. "The Evolution of Altruistic Behavior." *American Naturalist* 97: 354–356.

———. 1970. "Selfish and Spiteful Behavior in an Evolutionary Model." *Nature* 228: 1218–1220.
———. 1996a. "Spite and Price." In *Narrow Roads of Gene Land: Collected Papers.* vol. 1, *Evolution of Social Behavior,* pp. 171–184. New York: W. H. Freeman.
———. 1996b. *Narrow Roads of Gene Land: Collected Papers.* vol. 1, *Evolution of Social Behavior.* New York: W. H. Freeman.
Hammerstein, P., ed. 2003. *The Genetic and Cultural Evolution of Cooperation.* Cambridge, Mass.: MIT Press.
Hammett, Dashiell. 1989. *The Big Knockover.* New York: Vintage.
———. 1999. *The Maltese Falcon.* In *Complete Novels.* New York: Literary Classics of the United States.
Hardin, G. 1968. "The Tragedy of the Commons." *Science* 162: 1243–48.
Harding, D. W. 1962. "Psychological Processes in the Reading of Fiction." *British Journal of Aesthetics* 2: 133–147.
Harington, Joseph E., Jr. 1989. "Non-cooperative Games." In *The New Palgrave: Game Theory,* ed. John Eatwell, Murry Milgate, and Peter Newman, pp. 178–184. New York: W. W. Norton.
Hauser, Marc D. 1996. *The Evolution of Communication.* Cambridge, Mass.: MIT Press.
———. 2006. *Moral Minds: How Nature Designed Our Universal Sense of Right and Wrong.* New York: Ecco Press.
Hawkes, Kristen. 1991. "Showing Off: Tests of an Hypothesis about Men's Foraging Goals." *Ethology and Sociobiology* 12: 29–54.
———. 1993. "Why Hunter-Gatherers Work: An Ancient Version of the Problem of Public Goods." *Current Anthropology* 34: 341–361.
Hawthorne, Nathaniel. 1913. *The House of the Seven Gables.* Boston: Houghton-Mifflin.
Hazlitt, William. 1913. "On the Feeling of Immortality in Youth." In *Miscellaneous Essays.* London: Everyman's Library.
Hegel, G. W. F. 1977. *Phenomenology of Spirit,* trans. Arnold V. Miller. New York: Oxford University Press.
Henrich, Joseph, and Robert Boyd. 2001. "Why People Punish Defectors: Weak Conformist Transmission Can Stabilize Costly Enforcement of Norms in Cooperative Dilemmas." *Journal of Theoretical Biology* 208: 79–89.
Henrich, Joseph, et al. 2006. "Costly Punishment across Human Societies." *Science* 312(5781): 1767–1770.
Hitchcock, Alfred, dir. 1987. *Dial M for Murder.* Burbank, Calif.: Warner Home Video (VHS).
———. 1988. *North by Northwest.* Los Angeles: Janus Films (Videodisc).
———. 1990. *Notorious.* Los Angeles: Janus Films (Videodisc).

———. 1998. *The Lady Vanishes*. New York: Criterion (DVD).

Hoffman, Martin. 1981. "Is Altruism Part of Human Nature." *Journal of Personality and Social Psychology* 40: 121–137.

———. 1990. "Empathy and Justice Motivation." *Motivation and Emotion* 14(2): 151–172.

Homer. 1990. *The Iliad,* trans. Robert Fagles. New York: Viking.

———. 1996. *The Odyssey,* trans. Robert Fagles. New York: Viking.

Hume, David. 1975. *Enquiries Concerning Human Understanding and Concerning the Principles of Morals,* ed. L.A. Selby-Bigge. 3rd ed. Oxford: Clarendon Press.

———. 1978. *A Treatise on Human Nature,* ed. L.A. Selby-Bigge. 2nd ed. New York: Oxford University Press.

———. 2006. *An Enquiry Concerning the Principles of Morals: A Critical Edition,* ed. Tom L. Beauchamp. New York: Oxford University Press.

Hyde, Lewis. 1983. *The Gift: Imagination and the Erotic Life of Property.* New York: Random House.

James, Henry. 1937. *The Wings of the Dove.* New York: Modern Library.

———. 1956. *The Bostonians.* New York: Modern Library.

———. 1909. *The Ambassadors.* New York: Scribner's.

———. 1966. *The Turn of the Screw: An Authoritative Text, Backgrounds and Sources, Essays in Criticism.* New York: W. W. Norton.

———. 1987. *The Golden Bowl.* New York: Penguin.

———. 1999. *The Portrait of a Lady.* New York: Oxford University Press.

James, William. 1983. *Principles of Psychology.* Cambridge, Mass.: Harvard University Press.

Johnson, Samuel. 1759. *The Idler,* no. 40. Reprinted in *The Works of Samuel Johnson,* London: W. Baynes and Son (1824), pp. 414–148.

———. 1750. *The Rambler,* sections 1–54; from *The Works of Samuel Johnson, in Sixteen Volumes.* vol. 1. Electronic Text Center, University of Virginia Library. Available at http:etext.lib.virginia.edu/toc/modeng/public/Joh1Ram.html.

———. 1765. *The Preface to Shakespeare, Together with Selected Notes on Some of the Plays.* Available at http:etext.library.adelaide.edu.au/j/johnson/samuel/preface/.

Joyce, James. 1986. *Ulysses: The Corrected Text,* ed. Hans Gabler et al. New York: Random House.

Kafka, Franz. 1983. *Complete Stories,* ed. Nahum Glatzer. New York: Schocken Books.

Kaplan, H., and K. Hill. 1985a. "Food Sharing among Aché Foragers: Tests of Explanatory Hypotheses." *Current Anthropology* 26: 223–246.

———. 1985b. "Hunting Ability and Reproductive Success among Male Aché Foragers: Preliminary Results." *Current Anthropology* 26: 131–133.

Kavanagh, Etta, ed. 2006. "Debating Sexual Selection and Mating Strategies" [Letters on Roughgarden et al., 2006], *Science* 312 (May 5): 689–697.

Kleist, Heinrich von. 1810. "On the Marionete Theater," trans. Idris Perry. Available at www.southerncrossreview.org/9/kleist.htm.

Knoch, Daria, et al. 2006. "Diminishing Reciprocal Fairness by Disrupting the Right Prefrontal Cortex." *Science,* October 5 preprint (*Sciencexpress,* pp. 1–6).

Koch, Cristof. 2004. *The Quest for Consciousness: A Neurobiological Approach.* Denver: Roberts and Co.

Kreps, David M. 1989. "Nash Equilibrium." In *The New Palgrave: Game Theory,* ed. John Eatwell, Murry Milgate, and Peter Newman, pp. 167–177. New York: W. W. Norton.

Kreps, David M., and Joel Sobel. 1994. "Signaling." In *Handbook of Game Theory,* ed. R. J. Aumann and S. Hart. New York: Elsevier, 2: 849–867.

Kruger, Daniel J., Maryanne Fisher, and Ian Jobling. 2005. "Proper Hero Dads and Dark Hero Cads: Alternate Mating Strategies Exemplified in British Romantic Literature." In *The Literary Animal: Evolution and the Nature of Narrative,* ed. Jonathan Gottschall and David Sloan Wilson, pp. 225–243. Evanston, Ill.: Northwestern University Press.

Kubovy, Michael. 1988. *The Psychology of Perspective and Renaissance Art.* New York: Cambridge University Press.

Kurzban, Robert, and Daniel Houser. 2005. "Experiments Investigating Cooperative Types in Humans: A Complement to Evolutionary Theory and Simulations." *PNAS* 102(5): 1803–1807.

Kurzban, Robert, Peter DeScioli, and Erin O'Brien. 2006. "Audience Effects on Moralistic Punishment." *Evolution and Human Behavior,* Avaiable at http: digitalcommons.unl.edu/politicalsciencehendricks/7.

Kyd, Thomas. 1999. *The Spanish Tragedy.* In *Renaissance Drama: An Anthology of Plays and Entertainments,* ed. Arthur F. Kinney. Malden, Mass.: Blackwell.

Leiris, Michel. 1938. *Miroir de la Tauromachie.* Paris: Mano.

———. 1946. *L'âge d'homme; précédé de la littérature considérée comme une tauromachie.* Paris: Gallimard.

Levi-Strauss, Claude. 1976. *Structural Anthropology.* Chicago: University of Chicago Press.

Lewis, David. 1986. "Prisoners' Dilemma Is a Newcomb Problem." In his *Philosophical Papers.* vol. 2, pp. 299–304. New York: Oxford University Press.

Lewontin, Richard C., Steven Rose, and Leon J. Kumin. 1984. *Not in our genes: Biology, Ideology, and Human Nature.* New York: Pantheon.

Loftus, Elizabeth. 1996. *Eyewitness Testimony.* Cambridge, Mass.: Harvard University Press.

Lotem, Arnon, Michael A. Fishman, and Lewi Stone. 2003. "From Reciprocity to

Unconditional Altruism through Signaling Benefits." *Proceedings of the Royal Society* B 270: 199–205.

Mackenzie, George. 1660. *Aretina; or The Serious Romance.* Edinburgh: Robert Brown. Available at www.gateway.proquest.com.

Machado de Assis, Joaquim. 1998. *Dom Casmurro,* trans. John Gledson. New York: Oxford University Press.

Marías, Javier. 2005. *Your Face Tomorrow,* trans. Margaret Jull Costa. New York: New Directions.

Mauss, Marcel. 1990. *The Gift: The Form and Reason for Exchange in Archaic Societies,* trans. W. D. Halls. New York: W. W. Norton.

McCarthy, Cormac. 1998. *Cities of the Plain.* New York: Knopf.

———. 2005. *No Country for Old Men.* New York: Knopf.

———. 2006. *The Road.* New York: Knopf.

McGinn, Colin. 2004. *Mindsight: Image, Dream, Meaning.* Cambridge, Mass: Harvard University Press.

McKaye, K. R. 1991. "Sexual Selection and the Evolution of the Cichlid Fishes of Lake Malawi." In *Cichlid Fishes' Behaviour, Ecology and Evolution,* ed. M. H. A. Keenleyside, pp. 241–257. London: Chapman and Hall.

Melville, Herman. 1967. *Moby Dick: An Authoritative Text,* ed. Hayford and Parker. New York: W. W. Norton.

Mendes, Sam, dir. 2000. *American Beauty.* Universal City, Calif.: Dreamworks (DVD).

Merrill, James. 1982. *The Changing Light at Sandover: Including the Whole of the Book of Ephraim, Mirabell's Books of Number, Scripts for the Pageant, and a New Coda, the Higher Keys.* New York: Atheneum.

Miller, D. A. 1988. *The Novel and the Police.* Berkeley: University of California Press.

Miller, Geoffrey. 2000. *The Mating Mind: How Sexual Choice Shaped the Evolution of Human Nature.* New York: Doubleday.

Milton, John. 1975. *Paradise Lost: An Authoritative Text, Backgrounds and Sources, Criticism,* ed. Scott Elledge. New York: W. W. Norton.

———. 1998. *Paradise Regained.* In *The Riverside Milton.* New York: Houghton Mifflin.

Mithen, Steven. 1999. *The Prehistory of Mind: The Cognitive Origins of Art and Science.* New York: Thames and Hudson.

Moliére. 1965. *Œuvres Complètes,* volume 3. Paris: Garnier-Flammarion.

Moore, G. E. 2004. *Commonplace Book: 1919–1953.* London: Routledge

Moran, Richard. 1994. "The Expression of Feeling in Imagination." *Philosophical Review* 103(1): 75–106.

Morrison, Paul. 1996. "Noble Deeds and the Secret Singularity: *Hamlet* and *Phedre.*" In *Reading the Renaissance: Culture, Poetics, and Drama,* ed. Jonathan Hart, pp. 179–202. New York: Garland Press.

Nabokov, Vladimir Vladimirovich. 1958. *Lolita.* New York: Putnam.

Nagel, Thomas. 1974. "What Is It Like to Be a Bat?" *Philosophical Review* 83: 43–50.

———. 1986. *The View from Nowhere.* New York: Oxford University Press.

Nash, John. 2001a. "Equilibrium Points in N-Person Games." In *Readings in Games and Information,* ed. Eric Rasmussen, p. 9. Malden, Mass.: Blackwell.

———. 2001b. "Non-Cooperative Games." In *Readings in Games and Information,* ed. Eric Rasmussen, pp. 10–20. Malden, Mass.: Blackwell.

Nashe, Thomas. 1958. *Works,* ed. Ronald B. McKerrow. Oxford: Blackwell.

Nettle, Daniel. 2005. "What Happens in *Hamlet*? Exploring the Psychological Foundations of Drama." In *The Literary Animal: Evolution and the Nature of Narrative,* ed. Jonathan Gottschall and David Sloan Wilson, pp. 56–75. Evanston, Ill.: Northwestern University Press.

Von Neumann, John, and Oskar Morgenstern. 2004. *Theory of Games and Economic Behavior.* Princeton, N.J.: Princeton University Press.

Nowak, Martin A., Karen M. Page, and Karl Sigmund. 2000. "Fairness versus Reason in the Ultimatum Game." *Science* 289: 1773–1775.

Nuttall, A. D. 1996. *Why Does Tragedy Give Pleasure?* New York: Oxford University Press.

O'Farrell, Mary Ann. 1997. *Telling Complexions: The Nineteenth-Century English Novel and the Blush.* Durham, N.C.: Duke University Press.

O'Gorman, Rick, David Sloan Wilson, and Ralph R. Miller. 2005. "Altruistic Punishing and Helping Differ in Sensitivity to Relatedness, Friendship, and Future Interactions." *Evolution and Human Behavior* 26: 375–387.

The Onion. 2001. "Hijackers Surprised to Find Selves in Hell." Available at www.theonion.com/content/node/38673.

Ovid. 1916. *Metamorphoses,* ed. Frank Justus Miller. New York: Putnam.

Ozick, Cynthia. 1990. "The Shawl," in New York: Vintage.

Pinker, Steven. 1997. *How the Mind Works.* New York: W. W. Norton.

Pischedda, A., and A. K. Chippindale. 2006. "Intralocus Sexual Conflict Diminishes the Benefits of Sexual Selection." *PLoS Biology,* November, preprint.

Plato. 1997. *Complete Works,* ed. John M. Cooper and D. S. Hutchinson. New York: Pantheon.

Pollock, George H., ed. 1993. *Pivotal Papers on Identification.* Madison, Wisc.: International Universities Press.

Premack, D., and G. Woodruff. 1978. "Does the Chimpanzee Have a Theory of Mind?" *Behavioral and Brain Sciences* 1(4): 515–526.

Proust, Marcel. 1987. *A la recherché du temps perdu.* Paris: Gallimard.

Pynchon, Thomas. 1997. *Mason & Dixon.* New York: Henry Holt.

De Quervain, Dominique J. F., et al. 2004. "The Neural Basis of Altruistic Punishment." *Science* 305: 1254–1258.

Radnitzky, Gerald, and W. W. Bartley III, eds. 1987. *Evolutionary Epistemology, Rationality, and the Sociology of Knowledge*. La Salle, Ill.: Open Court.

Rasmussen, Eric, ed. 2001. *Readings in Games and Information*. Malden, Mass.: Blackwell.

Renoir, Jean, dir. 2003. *Le Crime de Monsieur Lange*. Paris: StudioCanal (DVD).

Richardson, Samuel. 1932. *Clarissa, or the History of a Young Lady*. London: J. M. Dent.

Richerson, Peter J. and Robert Boyd. 2005. *Not by Genes Alone: How Culture Transformed Human Evolution*. Chicago: University of Chicago Press.

Richerson, Peter J., Robert T. Boyd, and Joseph Henrich. 2003. "The Cultural Evolution of Human Cooperation." In *The Genetic and Cultural Evolution of Cooperation*, ed. P. Hammerstein, pp. 357–388. Cambridge, Mass.: MIT Press.

Riolo, R., Michael D. Cohen, and Robert Axelrod. 2001. "Evolution of Cooperation without Reciprocity." *Nature* 6862: 441–442.

Rosenberg, Marvin. 1972. *The Masks of King Lear*. Berkeley: University of California Press.

Roth, Philip. 2004. *The Plot against America*. Boston: Houghton Mifflin.

Roughgarden, Joan, Meeko Oishi, and Erol Akçay. 2006. "Reproductive Social Behavior: Cooperative Games to Replace Sexual Selection." *Science* 311: 965–969.

Rowell, Jonathan T., Stephen P. Ellner, and H. Kern Reeve. 2006. "Why Animals Lie: How Dishonesty and Belief Can Coexist in a Signaling System." *American Naturalist* 168: E180–E204.

Rowling, J. K. 1998. *Harry Potter and the Sorcerer's Stone*. New York: A. A. Levine.

Salmon, Catherine. 2005. "Crossing the Abyss: Erotica and the Intersection of Evolutionary Psychology and Literary Studies." In *The Literary Animal: Evolution and the Nature of Narrative*, ed. Jonathan Gottschall and David Sloan Wilson, pp. 244–257. Evanston, Ill.: Northwestern University Press.

Schelling, Thomas. 1969. "Some Thoughts on the Relevance of Game Theory to the Analysis of Ethical Systems." In *Game Theory in the Behavioral Sciences*, ed. Ira R. Buchler and Hugo G. Nutini, pp. 45–60. Pittsburgh: University of Pittsburgh Press.

———. 1978. "Altruism, Meanness, and Other Potentially Strategic Behaviors." *American Economic Review* 68: 229–230.

———. 2006. "Strategies of Commitment." In *Strategies of Commitment and Other Essays*, pp. 1–24. Cambridge, Mass.: Harvard University Press.

Scientific American. 2006. "Most Desirable Mates May Not Sire Most Prolific Offspring." *Science News. ScientificAmerican.com*, October 24.

Scott, Walter, Sir. 1960. "Proud Masie." In *Collected Works of Sir Walter Scott*. New York: Greystone Press.

Seabright, Paul. 2004. *The Company of Strangers: A Natural History of Economic Life*. Princeton, N.J.: Princeton University Press.
Searcy, William A., and Stephen Nowicki. 2005. *The Evolution of Animal Communication*. Princeton, N.J.: Princeton University Press.
Shakespeare, William. 2001. *The Arden Shakespeare Complete Works,* ed. Richard Proudfoot, Ann Thompson, and David Scott Kastan. New York: Arden Shakespeare.
Shaviro, Steven. 2003. *Connected, or, What It Means to Live in the Network Society.* Minneapolis: University of Minnesota Press.
Sidney, Sir Philip. 1971. *A Defence of Poetry.* New York: Oxford University Press.
Sigmund, Karl, and Christopher Hauert. 2002. "Altruism." *Current Biology* 12(8): 270–272.
Smith, Adam. 2004. *The Theory of Moral Sentiments*. Kila, Mont.: Kessinger
Sober, Elliott, and David Sloan Wilson. 1998. *Unto Others: The Evolution and Psychology of Unselfish Behavior.* Cambridge, Mass.: Harvard University Press.
Stanford, Craig B. 2001. *The Hunting Apes: Meat Eating and the Origins of Human Behavior.* Princeton, N.J.: Princeton University Press.
Storey, Robert. 1996. *Mimesis and the Human Animal: On the Biogenetic Foundations of Literary Representation*. Evanston, Ill.: Northwestern University Press.
Strawson, P. F. 2003. "Freedom and Resentment." In *Free Will,* ed. Gary Watson, pp. 72–93. 2nd ed. New York: Oxford University Press.
Sugiyama, Michelle Scalise. 2001. "Narrative Theory and Function: Why Evolution Matters." *Philosophy and Literature* 25: 233–250.
Thackeray, William. 2003. *Vanity Fair: A Novel without a Hero*. New York: Penguin.
Tooby, John, and Leda Cosmides. 2001. "Does Beauty Build Adapted Minds? Towards an Evolutionary Theory of Aesthetics, Fiction, and the Arts." *SubStance* 94–95: 6–27.
Trivers, Robert. 1971. "The Evolution of Reciprocal Altruism." *Quarterly Review of Biology* 46: 35–57.
Trollope, Anthony. 1879. *Thackeray.* London: Macmillan.
———. 1883. *Autobiography.* Leipzig: Tauchnitz.
———. 1973a. *Can You Forgive Her?* New York: Oxford University Press.
———. 1973b. *The Duke's Children*. New York: Oxford University Press.
———. 1983. *The Prime Minister.* New York: Oxford University Press.
Tucker, Albert. 2001. "A Two-Person Dilemma." In *Readings in Games and Information,* ed. Eric Rasmussen, pp. 7–8. Malden, Mass.: Blackwell.
Turner, Victor. 1982. *From Ritual to Theatre: The Human Seriousness of Play.* New York: PAJ Publications.

Turow, Scott. 1987. *Presumed Innocent.* New York: Farrar, Straus, & Giroux.
De Waal, Frans B. M. 2005. "How Animals Do Business." *Scientific American* (April): 73–78.
Walton, Kendall L. 1990. *Mimesis as Make-believe: On the Foundations of the Representational Arts.* Cambridge, Mass.: Harvard University Press.
Wedekind, Claus, and Manfred Milinski. 2000. "Cooperation through Image Scoring in Humans." *Science* 288: 850–852.
Wenseleers, Tom, and Francis L. W. Ratnieks. 2006. "Enforced Altruism in Insect Societies." *Nature* 444 (November 2): 50.
Wharton, Edith. 1994. *The House of Mirth.* New York: Oxford University Press.
Wilder, Billy, dir. 2002. Sunset Blvd. Hollywood, Calif.: Paramount (DVD).
Wilson, D. S. 2003. *Darwin's Cathedral.* Chicago: University of Chicago Press.
Wilson, E. O. 1975. *Sociobiology: The New Synthesis,* Cambridge, Mass.: Harvard University Press.
———. 1998. *Consilience: The Unity of Knowledge.* New York: Knopf.
Wittgenstein, Ludwig. 1967. *Zettel,* ed. G. E. M. Anscombe and G. H. von Wright. Berkeley: University of California Press.
———. 1997. *Philosophical Investigations,* trans G. E. M. Anscombe. Malden, Mass.: Blackwell.
Wollheim, Richard. 1974. "Identification and Imagination." In *Freud: A Collection of Critical Essays,* ed. Richard Wollheim, pp. 175–192. Garden City, N.Y.: Anchor Books.
Wood, B., and K. Hill. 2000. "A Test of the 'Showing-Off' Hypothesis with Ache Hunters." *Current Anthropology* 41: 124–125.
Woolf, Virginia. 1925. *Mrs. Dalloway.* New York: Harcourt, Brace, and Co.
Wouk, Herman. 1951. *The Caine Mutiny: A Novel of World War II.* Garden City, N.Y.: Doubleday.
Yamagishi, Toshio. 1986. "The Provision of a Sanctioning System as a Public Good." *Journal of Personality and Social Psychology* 51: 110–116.
Zahavi, Amotz and Avishag Zahavi. 1997. *The Handicap Principle: A Missing Piece of Darwin's Puzzle.* New York: Oxford University Press.
Zunshine, Lisa. 2006. *Why We Read Fiction: Theory of Mind and the Novel.* Columbus: Ohio State University Press.

Notes

Introduction

1. Although the Zahavis are skeptical of group selection I think (as they do not) that their argument is in fact consistent with the explicitly group-selectional argument of Sober and Wilson (1998); ultimately this may come to a question of terminology. The Zahavis' argument is certainly consistent with Robert Frank's views in *Passion within Reason* (1988). Marc D. Hauser sees what is called the principle of "strong reciprocity" involved in altruistic punishment as essential to human morality. I am attracted to his attempt to reconcile Rawles and evolutionary biology. Hauser, however, denies an impulse to convert defectors through punishment, whereas (as will become clear) I think such an impulse to convert is an essential element, mechanism, and manifestation of vicarious experience. See Hauser (2006), pp. 81–82. Gray (2007) offers a judicious critique of Hauser, while pointing out the great advance he makes on Dawkin's view of the motives for genetically unselfish behavior.

1. How Could an Interest in Fiction Have Evolved?

1. The phrase "evolution of cooperation" is originally Axelrod and Hamilton's (1981) and gives Axelrod the title of his groundbreaking book (1984). Sober and Wilson (1998), however, object to the term "cooperation," preferring the more polemical "altruism." I have no objection to maintaining shades of differences between them for the purposes of the argument here.
2. In *Darwin's Cathedral* D. S. Wilson makes a roughly similar claim for the power of evolutionary argument to explain the development of human religion, which for him is neither a pure adaptation nor a pure spandrel-like by-product of human evolution. In his important summary claim in that book, in the chapter on forgiveness to which in some ways my chapter on vindication below might be regarded as a pendant, he states: "We have an

innate capacity for altruism, selfishness, retaliation, forgiveness, contrition, generosity, commitment, saintliness, and vengefulness" (Wilson 2003: 193–194).

3. Tooby and Cosmides (2001: 7) have a more restricted notion of the fictional as the nonveridical, which mistakes the phenomenon that requires explanation.
4. Joseph Carroll (2004) has been a tireless but also tiresome expositor of evolutionary literary theory. Taking his cue from the straightforward and unimaginative pseudohumanism of E. O. Wilson (1998), who has endorsed his work, he essentially argues that all adaptations were adapted for what they are used for; that we use literature to help us understand ourselves and other people in a way that helps us be social; and that this is the reason we needed to create it. But Carroll, and Wilson seem oddly enough to be counter-evolutionary in proposing a Freudian paradigm for literature: life is scary but magical thinking helps us cope with its scariness. I accept this as a fact but not as a reason for the development of literature, which (as I'll argue) recruits capacities evolved for quite other reasons. It's a simple and heartwarming story that he and Wilson tell, but not a gripping one. Carroll's general mode is one of stunning contempt for any argument that doesn't treat the use of some capacity as the reason that capacity evolved.
5. Bruner (1990) sees narrative as essential to the need to give an account of oneself in the family and social dynamic, and therefore as an essential part of human interaction. (Indeed, Storey [1996] makes interesting use of his work.) What I have to say is consistent with his argument but I wish to explore the motivation for interest in narratives about others, even those, like fictional characters, who would seem to be nothing to us as we are nothing to them.
6. See Sugiyama (2001: 233, 235). In common with most evolutionary psychologists of narrative, she sees narrative itself as meeting the "recurrent adaptive problem or problems to which it may be an effective solution" (p. 235). That solution is a kind of trade in information, with the consumption of information regarded as an end in itself. While this may be true, I am interested instead in a different aspect of narrative, one in which it doesn't serve the ends of its speakers or hearers, but instead appeals to features of the mind that insure cooperation rather than survival and reproductive success. One merit of my claim is that it covers narrative on all levels of sophistication, just so long as readers are emotionally involved. A lot of evolutionary psychology treats only the most rudimentary or basic narrative motifs. Evolutionary psychologists are good at explaining fairy tales, but are not as good at explaining *Mason & Dixon* (Pynchon 1997).

And it won't do to say that *Mason & Dixon* is founded on the elemental structures of fairy tales, because it is still Mason and Dixon, and their friendship, that I care about, in a way quite different from that in which I care about those other navigators of the forest, Hansel and Gretel. Evolutionary psychologists tend to pay for the explanations they give of rudimentary or universal narrative structure by being unable to account for the cultural specifics that can make narrative so powerful. How those cultural specifics can come to play such an important role is one of the questions I'll seek to answer. I should say that Brian Boyd's (2001) account of how narrative works is quite alert to our capacities to understand what others are doing in very sophisticated contexts.

7. I think this is true of Tooby and Cosmides (2001) as well, who effectively have a Kantian sense of the aesthetic as managing and therefore appealing to the self-regulation of the mental apparatus. (This is a part of Zunshine's [2006] constellation of arguments as well.) Much of what they say is Kant and Aristotle in the trappings of evolutionary psychology—which I am glad of, since I like any endorsement of Kant and Aristotle. But they do seem to assimilate narrative to aesthetic experience in general, which seems unlikely since even on their showing aesthetic experience is more about the experience of form than of content. They give no sense (except perhaps a weakly identificatory one) of why we care about fictional characters. Much of what they say of narrative experience as a kind of play-practice for assimilating information about the world, and how to handle it, can also be found in Nuttall's *Why Does Tragedy Give Pleasure?* (1996). But they don't explain why we care so much and grow so absorbed in what we see. They see the force of the question, but don't seem to give an adequate answer. For more on them, see Note 9.
8. Thus *Rambler* 60, October 13, 1750: "All Joy or Sorrow for the Happiness or Calamities of others is produced by an Act of the Imagination, that realises the Event however fictitious, or approximates it however remote, by placing us, for a Time, in the Condition of him whose Fortune we contemplate; so that we feel, while the Deception lasts, whatever Motions would be excited by the same Good or Evil happening to ourselves" (Johnson 1750). In the *Preface to Shakespeare* (1765), as we'll see, Johnson presents a more likely account of sympathy.
9. One of the best attempts to argue for an evolutionary basis for aesthetic experience, that offered by Tooby and Cosmides (2001: 9–10), parallels Walton (1990) in assimilating fiction and pretense to each other. I think that, in general, evolutionary accounts of aesthetic experience have not cut nature at the joints, and have not given a good description of the experience to be explained. I hope my claim that to understand narrative

you need to consider our absorption in the nonactual, rather than the nonveridical (which is how Tooby and Cosmides define fiction [2001: 7]) seems more plausible. Although they debunk what they call the "appetite for the true" model of how the human mind responds to information, they don't consider the "appetite for the nonactual" (as we could dub it) that I think is central. Our desire to embellish, and to believe embellishments, from Homer to Joe McGinnis and the *National Enquirer,* suggests that the distinction between what we know is nonveridical and what we hope might be true is not a bright one, and allows for the analysis that I will give here. I'll also say here that like many evolutionary psychologists, Tooby and Cosmides seem to distinguish too rigorously between the notions of by-product and adaptation, perhaps in an attempt to ward off any charge of being like Stephen Jay Gould. Gould and Lewontin's (1979) spandrels metaphor was developed by Gould (most fully in 2002) into a tireless argument for the interplay between adaptation and by-product, so that lucky by-products could become recruited for new adaptations. I have found convincing only those evolutionary psychology arguments that see such an interplay. But Gould has been a whipping boy for the project of evolutionary psychology (see in particular Joseph Carroll's mean-spirited and red-baiting attack [2004: 227–245]). On the ad hominem quality of these attacks on Gould, see Buller (2005: 5–6). His judicious and well-informed book is a corrective to the blinkered polemicism that seems to infect nearly everyone engaged in arguing about evolutionary psychology.

Among the most virulently infected is Robert Storey (1996), who in a lively, entertaining, stunningly vicious and unreliable screed praises Carroll (2004) handsomely (Carroll reciprocates in pure tit-for-tat fashion; I believe the term is "log-rolling"), and also red-baits Gould, dismissing him (Storey 1996: 212n15) as a "Marxist"—bizarrely citing as evidence the fact that he doesn't *always* disagree with E. O. Wilson—even accusing him of the Stalinist biological opinions of Lysenko (Storey 1996: 205), while quoting him in a way so misleading as to risk bordering on fraud. At that time Gould was decently alive and could choose not to rise to the bait; Carroll's attack came after his death. But both Storey and Carroll strike one as anxious to stay in the good graces of E. O. Wilson, whose work and opinions on human culture they praise with very little qualification; Storey's work will also be gratifying to the polemically anti-Freudian Frederic Crews (1999), who has alas become something of a decanal presence within the evolutionary study of narrative. I regret Storey's simplistic polemicism, because he does occasionally fall into careless habits of accuracy, and a little of what he wants to say is consistent with my own argument, or at least with Robert Frank's (1988), whom he does not cite. But Frank's arguments against the simple-minded application of Trivers's (1971) claims (like his argument against the falsifiable and falsified economic claims of Richard Posner) would be uncongenial to Storey, who has found in the notion of the selfish gene looking out after its doubles the key to all human mythologies. I use George Eliot's phrase advisedly: Storey is far more amiable to Jung than to Freud, and I'll concede that his version of Jung is an interesting one, though perhaps it helps to show the weakness of his argument as well. Jung for him is essentially the way to mediate between pressures placed on humans as a *species* in Paleolithic times and the way individual humans essentially instantiate collective knowledge or thought about what archetypical humans are like.

Part of the point of my book is to assert the reconcilability of a Darwinian perspective, one that accepts evolutionary origins and constraints on human mental processing, with the best insights of European philosophy, psychoanalysis, and literary theory. That a theory of how we think should dismiss so much intense thought as beneath notice seems hasty. But Storey jeers at all of these enterprises with the haughty certainty and arrogance of the new convert to true religion breaking the idols of the marketplace. I am reminded of this great passage in William James, in a critique of Spencer and others, in 1890: "None the less easily, however, when the evolutionary afflatus is upon them, do the very same writers leap over the breach whose flagrancy they are the foremost to announce, and talk as if mind grew out of body in a continuous way" (James 1983: 147). Storey is no Carlyle but ends his book with the adjuration "Close thy Deleuze. Open thy Darwin" (Storey 1996: 205). I would prefer for everyone, no matter what their critical predispositions, to read them both, and not to imagine that praise of one requires dispraising the other.

Returning to the questions of mental functioning and its interactions as broached by Tooby and Cosmides (2001) (and praised by Storey), I want to cite the fascinating and widely influential book *Prehistory of Mind* (1999), in which Steven Mithen writes, "Whereas for Early Humans the domains of hunting, toolmaking and socializing were quite separate, these [became] so integrated that it is impossible to characterize any single aspect of Modern Human behaviour as being located in just one of these domains" (174).

A recent collection of essays on evolution and the nature of narrative (Gottschall and Wilson 2005) provides a good cross section of the state of the field. There are very interesting essays in it, and some very subtle literary criticism in the less polemical ones, but none seems to me to explain the reason for our absorption in fiction. On the whole, they either treat our interest in narrative as mainly practice for dealing with the real world, perhaps a mode that compresses knowledge of that world very efficiently, or analyze its *content* in terms of biological motivation (see also David Barash and Nannelle Barash 2005). I don't disagree with many of the specific claims they make; they just don't have very much explanatory power with respect to the question of why fiction is so all-absorbing. Even when Brian Boyd argues against Cosmides and Tooby's adaptationist theory (Boyd 2005: 169–170), he still describes only the function of literature. I accept that it has that function, but how do we explain the explicit experience of fiction, an experience the function does not seem to require? (Boyd concedes that some of this is mysterious.)

As to content, I can indicate my sense of how far some of these explanations fall short by citing two of the most interesting essays in the book, those by Daniel Nettle and by Catherine Salmon (Gottschall and Wilson 2005). Nettle explains very concisely how evolutionary biology can explain the interactions that we track with interest in drama; his interest in social tracking is harmonious with my own. But he sees such tracking as building on our natural and evolved proclivity to assess how our biological competitors are doing. And he somewhat retreats from this insight by assuming that dramas that show their characters attempting to enhance their own fitness will be maximally appealing. But why would we root for a happy ending, then, if the protagonist is a competitor? Nettle moves from an account of the use of drama to an account of the various things various plays could mean without (as far as I can tell) showing why that meaning has that use. Characters may have the biological or genetic motivations he says they have, but why do we care about them and their fates? How does fiction interest us? Likewise, Salmon explains why males and females might have different attitudes toward pornography and romance:

> Males evolved a sexual psychology that makes low-cost sex with new women exciting to imagine and motivating to engage in. Pornotopia is a fantasy realm, made possible by evolutionary-novel technologies, in which impersonal sex with a succession of high mate-value women is the norm rather than the rare exception. Ancestral females . . . had nothing to gain and much to lose from engaging in impersonal sex with random strangers . . . and a great deal to gain from choosing their mates carefully. The romance novel is a tale of female mate choice in which the heroine will identify, win, and marry the hero, who embodies the characteristics that indicated high mate value during the course of human evolutionary history. (Gottschall and Wilson 2005: 248)

Both Salmon and Nettle beg the question of what pleasure there is to see *others* doing what we wish to be doing. Salmon's impersonal formulation in her description of "a fantasy realm . . . in which impersonal sex with a succession of . . . women is the norm" elides the real question: who is having that sex? It's not the consumer of pornography, it's the actor in the pornography. Who is marrying the hero in the romance novel? Not its reader but its heroine. Why should this give pleasure to readers and viewers? Kruger, Fisher, and Jobling beg the same question in the same volume in their interesting taxonomy of heroes and the motivations women might have for choosing among them: "the dark hero and proper hero in British Romantic literature of the late eighteenth and early nineteenth centuries respectively represent cad and dad mating strategies" (Gottschall and Wilson 2005: 237). No doubt. But what is the nature of the way they "represent?" What does it mean that they "exemplify" (as the article's title has it) mating strategies? Why does literature do that? Why care with whom a heroine not oneself mates? Why care what nonexistent hero a heroine who does not exist selects?

10. Neuroscientists have recently been searching for what they call "mirror neurons," brain cells that respond alike both when I see and when I perform an action (see Gallese and Goldman [1998]). For a skeptical account of this modernized linkage of imitation and identificiation, see Gopnik (2007).
11. Wollheim (1974) begins with Freud on Leonardo's identification with his absent mother, which takes the form of supplying that absence through imitation. That sort of so-called identification is one that I accept completely. A fictional example may be found in Proust's description of how the young narrator of *A la recherché du temps perdu* attempts to make himself bald, because the charismatic Swann is and he wants Swann's love. He wants to be Swann as well. But he doesn't want to enter into Swann's thread of life (to cite a phrase Wollheim uses elsewhere); he wants to be Swann within his own life, that is, to bring Swann into his life. Identification as imitation of a lost or desired object takes the form of attempting to make *oneself*, one's own true biographical self, the object of the experience originally had of another, to make oneself the object of a love originally directed toward another. I would call it the converse of primary narcissism.
12. Harding (1962) and N. Carroll (1990; 1998) give strong critiques of the meaningfulness of the notion of identification.
13. See Hoffman: "Humans may be built in such a way that what happens to others is at times as motivationally significant as what happens to themselves" (1981: 127). Elsewhere Hoffman formulates the primordial experience of empathy as the real-life equivalent of Rawls's "veil of ignorance" in *A Theory of Justice* (1990).
14. Walton (1990) and Freud (1907). Even Wollheim (1974) says something

similar in his remarks on empathy and sympathy. For a useful correction, see McGinn (2004: 93).

15. N. Carroll (1990: 68–79) has an extended and convincing argument against Walton (1990), and especially this aspect of his theory.
16. See Moran's (1994) critique of Walton (1990).
17. Freud in particular and psychoanalysis in general, when not attempting to describe the psychology of literary response, don't treat identification naively, but rather as a propensity to imitate or internalize a positive or negative model, not in order to imagine oneself as subjectively *identical* to the object identified with but as a way to contain (or destroy) it within one's own internal world. Freud analogizes it to the oral-cannibalistic phase, whereby we take in but also potentially destroy the object we take in. I have nothing to object to Freud's formulation, in *Group Psychology and the Analysis of the Ego* (1921), that in "identifications the ego sometimes copies the person who is not loved and sometimes the one who is loved" (p. 107). Thus too does the narrator copy Swann, whom he loves, by trying to make himself bald like Swann. But in these cases it's not that we believe that we *are* the object of identification; rather, we take a vicarious relation to ourselves, treating ourselves as the objects we love or hate and playing in one person many people. This is why it's perfectly appropriate to relate projective identification (in which we seek to expel certain aspects of ourselves) to Freudian identification: they are essentially the same thing (if colored differently emotionally) in that they treat a part of the ego or self as an object to which we relate vicariously. See also Pollock's (1993) helpful anthology. Harry Frankfurt's (2004) account of identification with those aspects of our own affect, emotion, or desire that we wish to embrace or affirm also treats even aspects of the self, at least those we wish to interact with the outside world, vicariously.
18. See Harding (1962) and N. Carroll (1990; 1998). Walton (1990) gives an argument that might allow for a combination of these two aspects of imitation, since he argues that the spectator of fictional events or actions or configurations makes believe that he or she is seeing them. I doubt this, for reasons that I will go into below. See also Moran (1994). Ovid has Achaemides describe his anxiety on behalf of others when he sees Polyphemus throwing rocks at Ulysses's ship, forgetting that he is not on it: "Pertimui, iam me non esse oblitus in illa." (Ovid 1916: 14.186). He is, of course, describing the experience of anxiety that fiction produces in us on behalf of others, and taking issue with Lucretius's sense of the sweetness of our safety while watching from shore a ship in trouble upon the sea, in the famous tag "Suave mari magno" (which I think Shakespeare is thinking about at the beginning of *The Tempest*). I agree with Ovid: we *forget* that

we're not on the ship; we don't imagine or pretend or make believe that we are.

19. Premack and Woodruff (1978); Gopnik, Meltzoff, and Kuhl (1999) also argue explicitly against Descartes and in *The Scientist in the Crib,* rather like Wittgenstein (1997) and Austin (1970), that we derive a sense of our own minds vicariously from others (p. 47). This is a claim Gopnik developed earlier in an influential article (Gopnik, 1993). There she disavows Wittgensteinean (and Quinean) formulations, but is wrong to do so. Wittgenstein's description of human learning is very much like hers, though subtler. See also Zunshine on "'Effortless' Mind-Reading" (2006: 13–16). On gaze tracking in animals, see Zahavi (1997: 54–55). Gallese and Goldman (1998) disagree, but their evidence seems consistent with my claims. Their interpretation of this evidence seems to fall into the same conceptual error as Walton does.
20. Brian Boyd (2001) also talks about human theories of mind, our "folk psychology," and the ways that we track what others are doing: "Where other species 'read' other animals by differences in kind and by heuristics of position and behavior, humans read one another, and often other animals, with the help of a Theory of Mind, with a belief-desire-intention psychology" (p. 206).
21. Caillois (1987) would see following their actions as itself a kind of imitation; his extended idea of imitation is fine with me and allows me to think of our approaches as consonant with each other.
22. Anscombe (2000) notes that future intentions are couched in the present tense: I *am* going to do it. This would be because we get a description of a person turned toward the nonactual as though it were happening now. The same may be true of some attitudes toward the past: "I am afraid he did it." Moore's (2004: 291) much celebrated analysis of what's entailed by statements in the first person present indicative is especially relevant here. One can say "I believed it was raining but it wasn't," or "He believes that it's raining but it isn't," but normally not "I believe it's raining, but it isn't" (first person present indicative; this last is the formula Moore discusses). This means that present attitudes are different from past attitudes in their relation to belief and disbelief, which would sort with the immediate intensity that we bring to narratives. Though they are normally couched in the past tense—this is the central paradox—*our* attitudes to them are in the present. I believe (now) that Daniel will marry Gwendolyn, although the narrative in which it *turns* out he doesn't is in the past tense.
23. On the selfish gene see E. O. Wilson (1975), Hamilton (1970; 1996a), and Dawkins (2006).
24. Seabright, in *The Company of Strangers* (2004), is particularly good at syn-

thesizing the neurological, biological, and anthropological material on the mental capacities of primates, especially humans, and their ability to form large social units composed of unrelated individuals.

25. As I've noted, Caillois (1987) and perhaps Benjamin (1979) would probably call this imitation as well, but here I restrict the term to the way, say, Auerbach (1953) uses it. Caillois, who was among many other things a naturalist, and Jacques Lacan are important to any phenomenology of vicarious experience. This book is about biology, not phenomenology, but even here it is worth noting the connection between them. Donald T. Campbell, one of the founders of what he called "evolutionary epistemology" (Radnitzky and Bartley 1987: 47–89), discusses "vicarious locomotor devices" (56–58). The term is an apt one: vision evolves, he suggests, as a way of vicariously moving around by *looking* around and getting a vicarious sense of motion. This idea is anticipated in Caillois's sense of vision as itself a kind of imitation of the visual field by the seer.
26. See Fehr and Fischbacher (2003: 786–787).
27. Kurzban and Houser (2005: 1803–1807); R. H. Frank (1988).
28. Bowles, Choic, and Hopfensitzd (2003); Henrich and Boyd (2001).
29. Here I am following R. H. Frank (1988).
30. Axelrod (1984).
31. The Zahavis (1997); G. Miller (2000); Gintis et al. (2003).
32. See, e.g., Robert Frank's great book (1988); I give it as formulated by Tucker (2001) in 1950. The generalization of the Prisoner's Dilemma has been called "the tragedy of the commons" (Hardin 1968).
33. See Lewis (1986).
34. Basu (2007) suggests a similar analysis of the "Traveler's Dilemma" he formulated. Briefly schematized (and minus the narrative set-up that makes it fun), it goes like this: two players are asked to name a dollar amount between 2 and 100 simultaneously. They are told that whoever names the smaller amount will receive that amount plus $2, and whoever receives the larger amount will receive the *smaller* amount minus $2. If they name the same amount, they each get that amount. Most people name $100, so that if both name $100, both get $100 and do extremely well. But if I name $99 when you name $100 I'll get $101, so that's what I should do. And so should you. So I should name $98. And so should you, and so on. So that if we are both playing purely rationally, and with the expectation that the other is playing purely rationally, the purely rational choice of number to name (as can easily be seen) is $2. So instead of each of us getting $100, we'd each get $2, and both should losers be: rationality leads to loss.

    It turns out that no one, not even economists, plays Traveler's Dilemma rationally—since everyone recognizes that no one else will play it rationally. "What is interesting," writes Basu, "is that this rejection of formal rationality and logic has a kind of meta-rationality attached to it. If both players follow this meta-rational course, both will do well. The idea of behavior generated by rationally rejecting rational behavior is a hard one to formalize" (p. 95).

35. This is the main argument of R. H. Frank's book (1988).
36. Likewise, I might not defect if I thought you might harbor vindictive fantasies of revenge for twenty years, and then hunt me down and kill me, even at the cost of going to prison again for very little material benefit.
37. See Schelling (2006); R. H. Frank (1988).
38. Following Hamilton's (1970) account of spite, some biologists and psychologists have examined some of the parameters of spiteful behavior. One important parameter is what's called "image scoring." If you have a reputation for behaving badly, people will treat you spitefully (otherwise put, will tend to punish you altruistically). If your have a reputation for generosity, others will tend to behave altruistically toward you (Wedekind and Milinski 2000). In addition, if you have a reputation as an altruistic punisher, others will tend not to cheat around you. Put these all together and it's evident that a condition for the massive cooperation among nonrelatives to be found only in humans is that each individual's brain must be capable of keeping track of lots of reputations simultaneously. And each individual must have a "theory of mind" that attributes motivations to others and keeps score of those motivations.
39. One interesting way of faking or signaling irrationality that the originators of game theory (Von Neumann and Morgenstern 2004) consider is bluffing. On their account, almost certainly true, good players want to be caught bluffing from time to time, so that they'll get a reputation as bluffers, which will then serve them in good stead when they bet high on good hands.
40. I'll have more to say about "honest handicaps" later; the poker player's handicap is also an honest one in the sense that I will explore below, not for the negative reason that I analyze here, namely its artificial irrationality, but because his or her ability to win even by playing irrationally will spook other players when he or she bluffs later. On ways in which animals may bluff, see Rowell, Ellner, and Reeve (2006).
41. This is a very vexed notion (as some of the notes below will indicate). But I don't rely on a highly nuanced idea of "true altruism"—it is sufficient to my argument to describe it phenomenolistically as a disposition to behave in ways counter to a coldly rational calculation of one's best (reproductive) interests, even when one could get away with defecting.

42. The often-repeated story usually goes that one evening in a pub, when asked about whether he'd give up his life for someone else, Haldane replied, "I'd lay down my life for two brothers or eight cousins" (Quoted by Richerson and Boyd [2005]: 278n16).
43. The most famous argument for this widely accepted and well-understood phenomenon is Dawkins's *The Selfish Gene* (1976). George R. Price and Hamilton have shown that this model explains much of the mathematics of population genetics. See Hamilton's *Narrow Road of Gene Land* (1996b).
44. In the next chapter we'll have occasion to consider Bickerton's (2000) argument that human syntax derives from tit-for-tat altruism.
45. Alas, it strikes many cogent critics as the whole story. Fodor (2005) makes this mistake in a recent review in the *TLS* of Buller (2005) when he summarizes what he takes to be the Darwinian account of evolutionary psychology: "The 'ultimate' cause of [Jones's] behavior is his wanting to maximize his contribution to the gene pool" (p. 4). For Fodor, hostile as he rightly is to the kinds of arguments that Steven Pinker (1997) makes (see Foder, *The Mind Doesn't Work That Way* 2001), evolutionary psychology begs deep philosophical questions about human agency, action, desire, and motivation. On the argument I'll give, however, these philosophical questions remain alive and irreducible, since they are about *how* humans do think of other humans just precisely when there are good arguments within evolutionary psychology for *not* reducing human motivation to the maximization of the interests of one's own genes. If human motivation can't be reduced to the mechanical selfish-gene rule, then it occurs in an autonomous and independent region, where we can talk about human psychological characteristics that are helpful for sociability and social cooperation, without needing to see them as *designed* (in the way that Fodor suggests this word is question-begging) for individual fitness. Sociable beings do well in the world when they are involved with other sociable beings, but evolution didn't select for sociability. Rather, sociability turned out to be consistent with and to thrive under the conditions of the Environment of Evolutionary Adaptation. I would thus take Fodor's critique of evolutionary psychology as fairly well debunking the pure adaptationist claims made in literary studies by Joseph Carroll (2004), but not as debunking the more recent studies I cite here and that tend to undermine the "fundamentalist" positions of Dennett (1995) and Dawkins (2006). Hauser (2006: 82) also disagrees with Trivers (1971).
46. They are often discussed and are an essential part of the literature in accounts of behavior challenging rational choice theory. For a good account of the ultimatum game, see Güth and Van Damme (1998). Also Robert

Frank (1988). For accounts of strong reciprocity see Fehr and Fischbacher (2004).

47. Fehr and Fischbacher (2004), among others, list the relevant papers. See also Fehr and Henrich (2003: 4), where they treat purely anonymous sequential one-shot Prisoner's Dilemmas (in a sequential Prisoner's Dilemma the second player decides what to do based on what the first player has done). In those cases that present maximum incentives to defect, 40 to 60 percent of second players typically reward trust on the part of the first player, and 40 to 60 percent defect. Interestingly, in complicated multiperson games, when a receiver also controls the payoff to a third, passive, or silent receiver, the fact that the first receiver is put into a kind of dictator position as well interferes with his desire to ensure fairness to the silent receiver, even though the receiver might pay his own money to ensure fairness if he weren't also playing the game. Playing and spectatorship are therefore highly separate experiences, which again tends to disprove the identification theory of vicarious experience. We will return to this idea in the vindication chapter. See Camerer (2003: 80–83) and Güth and Van Damme (1998).

48. "'I can only *believe* that someone else is in pain, but I *know* if I am.' . . . Just try—in a real case—to doubt someone else's fear or pain." Wittgensttein (1969:102e). Gopnik (1993) relates such knowledge to perfectly practiced expertise.

49. Vicarious experience is genuine experience, open to nuanced consideration and surprise. In game theory there is a significant discrepancy between the way self-enforcing agreements actually arise and the way they *should* arise around Nash equilibria (Nash 2001a and b.) A Nash equilibrium (in a two-person, noncooperative game, where noncooperative is a technical term meaning that no institution enforces side agreements, commitments external to the game itself, to make some moves and not others) is a situation where each player can see the other's best strategy if he or she plays his or her best strategy, and can see that the other can see his or her best strategy. In these cases it should be clear to both what move to play. Such cases are often "self-enforcing," since each expects the situation to enforce the optimal move on the other, and therefore each makes the optimal move him- or herself. But in many situations A will rightly anticipate that B will anticipate that A may not make A's best move, for fear that B won't, in which case what would be B's best move when A makes his or her best move will be disastrous, and therefore A and B will both play nonoptimal moves. For a fascinating exposition see Kreps (1989) and Harington (1989). A significant element of human psychology enters the game, and what turns out to be the best move in such situations has to do with a re-

alization that other people are not or do not take one to be strictly rational. This discrepancy is significant to the accounts of strong reciprocity and altruism that I summarize here and to the kinds of ways we track how other people track still others or ourselves tracking them.

The most interesting and in some ways the most influential account of a closely related aspect of game theory for thinking about human ethics and morality is Thomas Schelling's. In "Some Thoughts on the Relevance of Game Theory to the Analysis of Ethical Systems," he considers, among much else, a counterpart of Prisoner's Dilemma called the Altruist's Dilemma (where both players attempt to be maximally altruistic and fail), on his way to the following culminating example:

> Take the familiar Prisoner's Dilemma matrix and suppose that both of us want to reward cooperators and to punish noncooperators; and suppose that that interest outweighs the personal payoffs reflected in the original numbers. The situation now has two equilibria. If we both cooperate, each is rewarded for his cooperation, each has rewarded the other for cooperating, indeed each has (properly) rewarded the other for rewarding onseself for being good . . . and we are happy all down the line. But if we both decline to cooperate, and "lose" in the usual sense, each of us is, superficially, content. I have punished a noncooperator by my refusal to cooperate. So has he. But this equilibrium lacks the infinite convergence of the other. I have punished him for refusal to cooperate; should I have rewarded him for punishing me for my refusal to cooperate? But then he failed to reward me for punishing a noncooperator, and perhaps deserves to be punished . . . But then if neither of us is really much concerned with the original numerical payoffs, nobody has either cooperated or failed to cooperate!
>
> I am not solving the problem, I am introducing it. The way game theory has managed to cope with the infinitely reflexive problems of "he thinks I think he thinks I think . . ." may provide a model for handling some interactions of motives. (1969: 59; his ellipses)

Brian Boyd notes the question of what to think about how other people will think about one's own thinking as a vivid feature of a human theory of mind: "The capacity to understand false belief . . . is crucial to the human capacity for narrative, and to narratives real and invented: Elizabeth Bennet *thinks* that Darcy *thinks* that she *thinks* he *thinks* too harshly of her family; Maria *foresees* that Sir Toby will eagerly *anticipate* that Olivia *judge* Malvolio absurdly impertinent to *suppose* that she *wishes* him to *regard* himself as her preferred suitor" (2001: 207). Zunshine makes a similar point with respect to a passage in *Mrs. Dalloway* (2006: 27–36). I think of one of the moments in *The Golden Bowl* when Maggie and Adam are aware of each other's awareness of each other: "He became aware himself, for that matter, during the minute Maggie stood there before speaking; and with the sense moreover of what he saw her see he had the sense of what she saw *him*" (James 1987: 149). The fascinating upshot of this will be the far later fact that we do not know—no one knows—what Adam is thinking at the end; he has achieved what the game theorists call perfection, since he need never make his move; the question whether he knows there's a move that makes itself through its pure potential to be made, preventing others from laying themselves open to it, is unanswerable, and he is inscrutable. The threat is enough, and he needn't know the threat is there. This is a self-enforcing equilibrium. Again I want to stress that the language of Nash equilibria is consonant with Schelling's and Robert Frank's related views about solving commitment problems, as this sentence in Strawson can also suggest: "if such a choice were possible, it would not necessarily be rational to choose to be more purely rational than we are" (2003: n4).

I should add one more technical point. The game theory that Hamilton (1970) and others have used to analyze evolutionary psychology is the theory of "noncooperative games." That term should not be allowed to be misleading. Noncooperative games are those in which cooperation can be self-enforcing, rather than volunteered by the cooperators. It seems unlikely that cooperative games, in which the players agree to help one another for the benefit of the whole group, could be evolved through Darwinian mechanisms. Purely voluntary cooperation against one's own interests would eventually die out. Nevertheless, a recent article by Roughgarden, Oishi, and Akçay (2006) has received a lot of attention in the press for its argument that cooperative game theory describes biological behavior more accurately than the standard (noncooperative) type. Their article is much ridiculed in scientific circles, to my mind justly (see, e.g., Kavanaugh, 2006), so I make no other reference to it. If they were correct it would probably help my argument, but they are almost certainly not correct. Indeed, the most embarrassing moment in their paper sounds to me as vapid as some of the evolutionary literary critics I take to task in these footnotes: "We hypothesize that a sense of friendship resides in animal bonding, a joy or synergy in the spirit of cooperation that allows animals to experience the product, and not merely the sum, of their individual well-beings" (p. 966). They give no evidence to back this hypothesis.

50. See the data graphed in De Quervain et al. (2004: 1255); their interpretation of the data is consistent with mine. See also Camerer (2003).
51. Dowd (1996) has a somewhat more subtle point than I'm crediting her

with, namely that charity for the sake of publicity distorts the process, since naming a university building at Harvard garners more publicity for the buck than giving the money to Oxfam for third-world well-digging. This is true, and is moreover a point that the Zahavis (1997) and Geoffrey Miller (2000) stress in their argument that altruism is a form of display, like the peacock's tale, an argument to which we'll return below. But from the giver's point of view, the donation is altruistic none the less, since the benefit in fame is certainly dwarfed by the cost.

52. The definition of spite is given by Hamilton, especially in his classic 1970 paper; the mathematics is owed to his conversations with George Price. See also Sigmund and Hauert: "Strong punishing behavior [when] the possibility of future benefits . . . is excluded . . . is usually called altruistic punishment, although in the strict sense it falls under the definition of spite" (2002: 270).
53. Egeus probably wouldn't count, since he is trying to *manage* his descendents, not prevent them.
54. Shylock expected the loan to be paid back, so he wasn't rationally calculating the benefit of killing Antonio from the start. It's worth noticing, as Hamilton (1970) would, that this benefit cannot possibly be worth the price he is willing to pay. Antonio cannot lend gratis (as Shylock complains) the sum of money that he now owes (which is why he has to borrow it), and even if he could, the loss to Shylock is only a fraction of the gain he would get from Antonio's paying him back, since the losses due to Antonio's interest-free loans have to be divided among *all* the money lenders, whereas the profit that Shylock gets from loaning Antonio money of the same order of magnitude goes to him alone. I have no doubt that this was obvious to Shakespeare, shrewd as he was at business.
55. Minding other people's business often takes the form of gossip. Brian Boyd (2001: 200) describes how, in comparison to other animals, humans are "more eager to keep track of one another; more attentive to the complex goals of others; much more capable of keeping track of others and their goals indirectly, through narrative, through gossip; more eager to share such information, since sharing it secures attention to the teller and offers information for the listener; and more likely to respond even to indirect reports with strong empathy or indignation." Axelrod lists gossip about violators of the social norm as a vehicle for their punishment in *The Complexity of Cooperation* (1997).
56. True, the selfish-gene theory can explain why the laws of sexual selection might make me act in ways that we would ascribe to vanity, why I might act like a very, very pajcock. But as we've seen, this theory can't by itself explain other passions, such as anger and the desire to punish an adversary.

57. Sober and Wilson (1998) give several plausible pathways by which this might happen. I doubt that any other footnote cites them and Sartre together, but I do so here: this is a way of saying that biology has brought humans to a place where genetic essence does not necessarily "precede human existence." It's not that humans are unpredictable, but that we manifest ways of behaving that are unexpected on the theory that the gene always seeks to maximize its reproductive success. Our genetic heritage reached (or "emerged at") a level of complexity that, by what Shelley calls a strange and natural antithesis, undermined the inflexible law that produced it, and made compromise more efficient than coldly rational maximization. I will offer more justification for this claim below.

    Ainslie (2001: 92–94) gives an illustration of what he calls "intertemporal bargaining" between present and future selves (such bargaining is a feature of addiction, for example, since my present self enjoys a cigarette my future self would prefer me not to smoke). His analysis can be extended to those genetic future selves represented by my offspring, whose potential existence (before they actually exist) my genes are much more interested in than I am.

58. It's significant that not only Zola but also so subtle a psychologist as Proust (1987, e.g. 2: 245–246) has no difficulty with describing people on both hereditary and experiential levels. We experience our heredity, and our heredity must take that into account. Even more than Zola, Proust sees perfect consistency in describing most specific historical and social commitments in what we could call genotypical terms.

59. They were looking at brain activity during variations of the ultimatum game. For a recent experiment also done in collaboration with Fehr on the part of the human brain that seems to suppress self-interest in favor of altruistic punishment during ultimatum games, see Knoch et al. (2006).

60. Dante's Virgil, in anatomizing the various forms of wishing evil on others, traces them all to love (*Purgatorio* 2003: 17.112–123). In particular, the desire for punishment or vengeance *(la vendetta)* is described as a mode of love, and the only way I can quite make sense of this is that Virgil sees even in vengeance and vindictiveness a desire to communicate with their object. The characteristic of such a desire is communicative anticipation: we look forward to the result of the experience we will communicate. I'll return to this below, in Chapter 4, on vindication and vindictiveness. Note that Freud is puzzled by something similar (which he couches in the language of identification I resist): "There is still much to be explained in the manifestations of existing identifications. These result among other things in a person limiting his aggressiveness toward those with whom he has identified himself, and in his sparing them and giving them help" (1921: n111).

61. I will explore this in some detail in my chapter on anticipations of vindication, where I'll be concerned with anticipatory gloating.
62. See my comments on this in *Generosity and the Limits of Authority* (1992: 43–44).
63. This is a sort of converse of what Bersani and Dutoit (1985) describe as the dysfunctional relation of sympathy, in which to empathize with others we fantasize their being in pain. Bersani and Dutoit derive this idea from the half-sympathy half-sadism of Assyrian representations of hunting, in which the pain of the wounded animal redounds to the prestige of the king wounding it.
64. Compare this to Anscombe's (2000) remarks on the present tense formulation of future actions, mentioned above.
65. Some people think that in modern, large-scale societies, cooperation will begin to be selected against. Trivers (1971) is the most prominent of theorists who think cooperation might be a stone-age relic and a maladaptation in the modern world. Trivers generally argues against the possibility of true altruism, since he sees competition everywhere, even among kin, even in single organisms. For arguments against this view see Fehr and Henrich (2003), Robert Frank (1988), and Sober and Wilson (1998).
66. I am primarily following the Zahavis (1997), Geoffrey Miller (2000), and Gintis, Smith, and Bowles (2001) in their arguments about how signaling benefits work (see below). On the central importance of the psychology, see Strawson: our "practices, and their reception, the reactions to them, really are expressions of our moral attitudes and not merely devices we calculatingly employ for regulative purposes" (2003: 93).
67. Below we'll consider more complex ways in which group selection can *derive* from the behavior of individuals.
68. It has just been discovered by Wenseleers and Ratnieks that a policing mechanism that makes defection a losing strategy ensures cooperation in some insect societies. They conclude: "The key role of relatedness in the evolution of self-sacrificing behaviour is widely recognized. The origin of insect societies is one of the most cited examples, and high relatedness was probably required for worker behaviour first to evolve. Nevertheless, our results show that in modern-day insect societies it is mainly social sanctions that reduce the numbers of workers that act selfishly" (2006: 50).
69. Axelrod and Hamilton (1981); R. Frank (1988); Richerson, Boyd, and Henrich (2003). Henrich et al. (2006) review the very consistent evidence for "costly punishment across human societies."
70. See Yamagishi (1986).
71. I have not found many experiments showing people altruistically rewarding altruistic punishers, or punishing those who fail to punish, as Axelrod

and others have modeled it. Henrich et al. do show that "costly punishment positively covaries with altruistic behavior across populations" (2006: 1767). An article by Gürerk, Irlenbusch, and Rockenbach (2006) shows that people will prefer, or come to prefer, to be members of groups in which altruistic punishment is the norm. Denant-Boèmont, Masclet, and Noussair show experimentally that people will sanction "not only free riders in the contribution phase, but those who deviate from group norms of punishment in a direction viewed as opportunistic" (2005: 21; see p. 15 and Table 7 as well). In general, such studies would probably be very difficult to design since the experimenter's necessary neutrality would itself look to subjects something like second-order free-riding. Another reason such studies would be particularly hard to set up is the thin line between behavior that we admire as altruistic punishment and the altruistic punishment gone-wrong that makes us particularly angry (as I'll claim in more detail below; this is an issue that comes up in Denant-Boèmont, Masclet, and Noussair, just cited). I do, however, offer the literary works I analyze as evidence for such a tendency to reward altruistic punishers: we spend time we could spend more productively (and reproductively) reading about and rooting on the good guys, the redressers of wrongs, cheering their successes and cursing their failures. The author of the book is the equivalent of the experimenter in the experiments I cite; the author is not neutral, but it's just the way he or she deviates from neutrality that makes us approve or disapprove of him or her. This doesn't mean all books have to end happily for us to approve; rather, the fates of characters have to seem worth it. Their sad endings, for example, should exemplify some version of true altruism and not just random failure. (I mean this about the satisfactions of narrative and of absorption in a fictional world; I'm not making a very strong claim here, since what I am saying is essentially that we like characters we have reasons to like.)

It's worth noticing at least one important paper that deals with fictional scenarios because it would be impossible to design an experiment which could match the phenomena being examined. O'Gorman, Wilson, and Miller (2005) compared responses to fictional scenarios about active helping and those involving altruistic punishment, and found that genetic relatedness matters far more to offering help to people than to punishing those who have wronged them. Altruistic punishment seems to engage us even when the victims of evil behavior are nothing to us. They noticed something very telling about their fictional scenario offering the imaginary possibility of altruistic punishment: "The altruistic punishment scenario strongly engaged the interest of the participants, as indicated by their reported degree of anger, their desire to see the transgressor punished, and the willingness of at least some of the participants to punish at their own expense. The variables of genetic relatedness, friendship, and potential for future interactions had no effect despite a strong overall psychological response" (2005:383). This emotional strong engagement came even though the subjects knew the scenario to which they were reporting their response was hypothetical and fictional.

In an e-mail to me (October 27, 2006) Joseph Henrich says that "there are a rash of papers coming out showing the punishment of non-punishers." One of them, Kurzban, DeScioli, and O'Brien (2006), shows that an anxiety about being seen not to punish a defector animates much altruistic punishment, just as Axelrod (1997) had predicted.

72. More likely punishment because instances of defection do more damage to cooperation than instances of cooperation reinforce it. Therefore an ESS will tend to elicit far more altruistic punishment than altruistic reward. This is again a version of the marginal asymmetry between actions whose costs seem equal, like that which Hamilton describes (1970). It is because of that asymmetry that it is better to save five siblings than kill twenty enemies.
73. In my account of altruism as a costly signal I have been strongly influenced by Geoffrey Miller (2000), and by Grafen (1990a), the Zahavis (1997), and Gintis (2000). Consider in particular this summary from the Zahavis: "A group whose members compete for prestige by demonstrating their altruism will be better equipped to compete against other groups than a coalition whose members vie for prestige by harassing those of lower rank or engaging in wasteful display" (1997: 149).
74. Robert Frank (1988: 23–24) and John Maynard Smith (e.g., Cronin and Maynard Smith 1993: 202–203) both treat this "attractive sons" argument, which derives from R. A. Fisher's classic book, *The Genetical Theory of Natural Selection* (1958: esp. 136–137), but is already implicit in his 1915 article. For an interesting and suggestive reading of Darwin's own account of sexual selection, see Grosz (2005: 64–92). But Grosz doesn't take up any of the later ideas, from Fisher to the Zahavis (1997), about the topic.
75. The Zahavis (1997: 39) give a powerful argument against the still prevalent idea that sexual and natural selection are independent of or in conflict with each other, pointing out that the same signals that attract mates deter rivals, which means the signals are not merely arbitrary or conventional; if they were, their costliness would mean that rivals would have a significant advantage in *natural* selection that would flood the effects of runaway sexual selection (see also Grafen [1990b]), which I describe at greater length below). The mediated desire implicit in the idea that animals select mates that will produce offspring attractive to other animals in the same species

can lead to the idea that they choose mates on the basis of their signaling (as opposed to what the Zahavis call utilitarian) features. But what the signals had better advertise *are* those utilitarian features. The Zahavis cite actual mediated desire among grouse, where younger females learn *which* males to choose by observing and imitating older ones (1997: 35), and I am aware of studies showing that goldfish similarly engage in imitative desire. This is significant because it suggests that some organisms have to learn to interpret the significance of signals, which means they are not genetically hardwired. Note that the Aristotilean idea of learning through imitation comes into play here. Again, though, identification with the organism they imitate is not the issue; rather, they track the tracking of other grouse and adjust their own goals accordingly.

76. See Mauss (1990); Bataille (1991); Flesch (1992).

77. Amotz and Avishag Zahavi (1997) don't mention Mauss (1990), but they seem to confirm his argument independently. They make the converse and Maussian point that "one can even view altruistic acts as implied or surrogate threats, since the prestige the altruist gains allows him or her to achieve what other animals gain by threats" (1997: 149). But they are *not* saying that the truth about altruism is that it's a veiled threat; rather, threat and altruism are both manifestations of the more basic category of costly signaling or assumed handicap.

    You can see one aspect of the cooperation between predator and prey in the doubling cube in backgammon. If you're winning a game it's always and obviously right to offer to double the stakes. If you're *losing* it's still sometimes right to accept the offer. If your chances of winning are greater than 25 percent, you should accept the double because your *expected* loss when the stakes are doubled is less (since you'll win over 25 percent of the time) than your *definite* loss of the original stake. Consider an original stake of $3, and let's say your chances of winning are reduced from 50–50 at the start to one-third (which is two-thirds of 50–50). Your opponent doubles. If you accept the double the stake is now $6, and you'll expect to win two-thirds of what you would have to win to break even. So you expect to lose $2 on the double. But if you refuse it you lose $3, the original stake. So you should accept the double. In this way it's to the advantage of both players to cooperate—to agree to play for double the stakes—given each player's alternative. I offer this as an illustration of how adversaries can best serve their opposing interests by cooperating in the structure of their contest. (This isn't an example of costly signaling; rather, costly or honest signaling is an example of such cooperation.)

78. In later editions of *The Selfish Gene* (originally published in 1976)

Dawkins adds a long footnote retracting his considerable skepticism and praising Amotz Zahavi's theory, and especially Grafen's (1990a) mathematical interpretation of Zahavi's argument. He concedes that if Zahavi is right, much of what he believes would have to be reassessed. This would include the argument made in Dawkins and Krebs (1978) that signalers manipulate their signals in order to cheat. Recent work has considered the tolerance for cheating or true bluffing within a system generally conforming to the handicap principle of honest signaling. See in particular Rowell, Ellner, and Reeve (2006) and Searcy and Nowicki (2005: 181–224).

79. See Flesch (1992). When Cleopatra describes Antony as "past the size of dreaming" she is making a sublime version of the argument that his charisma could not be faked, and must be an honest signal of his transcendent worth. I agree—or at least agree that only Shakespeare could dream up such a figure as Antony if he weren't real. (In doing so, and in having Cleopatra deny that Antony could have been an imaginative creation, Shakespeare is doing his own form of costly signaling.)

80. Boone (1998) considers food sharing from an anthropological point of view as a form of such costly signaling; see Stanford (2001), and also Hawkes's elegant study (1991: 37–38) of how adulterous women among Ache favor sharers of meat; she reviews the strong evidence that such sharing is indifferent to expectation of Trivers-like reciprocation in the future (for some caveats see Wood and Hill [2000] and Kaplan and Hill [1985a]). Bliege Bird, Smith, and Bird (2001) show how spear-fishing and turtle hunting function as very costly signals of fitness among the Meriam (inhabitants of the island of Mer in the Torres Strait, Australia). Game that is difficult to obtain is more widely shared, with far less advantage to the hunter and his family, than the same game when it is easier to obtain. Hunting is harder than collecting the same game, but hunters share more of what they obtain by hunting than do collectors by collecting, so that the hunt together with the distribution of the game functions as a costly and widely broadcast signal (Bliege Bird, Smith, and Bird 2001: 12). Compare the high likelihood today that fishers for sport or fly casters will share or give away their catch.

81. Compare Hotspur's insouciance, that is compare that signal in Hotspur that we (rightly) consider his generous insouciance:

> I do not care: I'll give thrice so much land
> To any well-deserving friend;
> But in the way of bargain, mark ye me,
> I'll cavil on the ninth part of a hair.
>
> (Shakespeare, *1 Henry IV* 2001: III.i.131–134)

82. Sober and Wilson (1998) review the powerful evidence for the existence of group selection throughout biology, in what they call multilevel selection. I said above that simple group selection is pretty much an exploded theory; however, the way it works in biology is not as simple as the way Darwin imagined it working, and in fact group selection *requires* dispersal of altruists at some point into other groups. The argument is elegant, beautiful, and supported by the facts, and many of the original and most important skeptics of group selection, especially Hamilton (1996a and b), have come around. Still, as Sober and Wilson say, most people are reflexively hostile to the idea, partly because they haven't considered it deeply, as Hamilton himself complained. Part of the problem is the definition of a *group,* an ill-defined notion in the debates until recently. Sober and Wilson give the simplest and most helpful: "In all cases, a group is defined as a set of individuals that influence each other's fitness with respect to a certain trait but not the fitness of those outside the group. Mathematically, the groups are represented by a frequency of a certain trait, and fitnesses are a function of this frequency" (1998: 92). I should note that the Zahavis (1997: 23–24), whom Sober and Wilson do not mention, don't accept an earlier and somewhat simpler version of group selection, since they think it would be too easy for cheaters to invade such a group. They consider a case where signaling could *replace* the quality being signaled. The arguments I am rehearsing here would meet their objections, and I think that Sober and Wilson would see them as being of the group selection party without knowing it, especially given their hostility to the selfish-gene notion of inclusive fitness (see the Zahavis 1997: 163–165). In an e-mail to me (November 17, 2005), however, Amotz Zahavi writes, "I think that Sober and Wilson are totally wrong. It is easy to explain phenomena by group selection when you have little data. Whenever I started being interested in explaining phenomena that are usually explained by the variety of indirect selection I found data that explain it by individual selection." This may be part of the hostility that Hamilton notes. At any rate, they too offer a formulation of the crucial insight about altruism on which I'm relying, in discussing babblers, a kind of bird they have observed for decades in the Negev and have studied assiduously: "The 'altruistic' individual thus serves its own interests—it expends efforts to demonstrate its quality reliably. Because it does so by engaging in altruism, rather than by pure showing off, as a peacock does with its tail and a bower bird with its bower, the babbler is also showing its interest in continued collaboration with the members of its group. Groupmates pay attention to the dual message because this information enables them to make their own decisions. At the same time, the whole group benefits from the actions that convey the altruist's message" (Zahavi 1997:157).

83. See Lotem, Fishman, and Stone (2003: 204): when "punishment acts as an effective negative reinforcement . . . the ability to punish can also signal the superior quality of the punisher."
84. Brian Boyd (2001: 206) rightly argues, "We have a default concern for others, and especially for others who stand out in terms of might and merit, and a default sympathy with their pursuit of their goals." He describes "the sort of agency we would naturally want to ally ourselves with" as "powerful, generous, and resourceful." He doesn't say why we have such a default interest, though, nor does he follow up on the idea of our natural desire to ally ourselves with such agents as I am seeking to do here.
85. De Waal (2005) describes the prototypes of such behavior in "chimpanzee economies," where he shows that chimps keep track of one anothers' behavior and reciprocate accordingly.
86. See Bliege Bird, Smith, and Bird (2001) on hunters' reputations as reported by narratives about them.
87. Grafen (1990b).
88. See Sober and Wilson on mutual aid among true altruists (1998: Chapter 2).
89. See Gintis, Smith, and Bowles (2001: esp. 113–115) on "signaling prosocial quality," and Bliege Bird, Smith, and Bird (2001).
90. See Lotem, Fishman, and Stone (2003), where they are called "additive signaling benefits."
91. See also Morrison (1996) on why Hamlet is fat and scant of breath.
92. See Gintis, Smith, and Bowles (2001), and also Guilford and Dawkins (1991).
93. Recall Hamilton (1970) and Sigmund and Hauert (2002) on spite.
94. In most experiments testing people's willingness to pay to punish, punishment costs them less than it costs their victims. Gintis (2000) reviews the evidence that tool-using humans made a strong breakthrough in low-cost punishment, since we can produce enormous harm at relatively low costs (by throwing rocks or shooting guns). Some experiments have tested the limits of altruistic punishment, or spite, in situations where, as with Malvolio, punishment hurts the punisher more than the victim (Fehr and Gächter 2000). Even in these situations, altruistic punishment is robust.
95. This leads to the kind of interesting self-reflective dynamic that I quoted Schelling on in note 49 above. See Grafen (1990b). In an e-mail to me (April 2, 2005), Grafen finds the extension and application of his analysis to humans, as I give it here and in note 49, reasonable: "I can see a logical possibility of the self-referentiality of the signal, if I can put it that way. In economics, with definitions of utility that made the utility of other individuals an argument in the function defining the utility of an individual, it

might well work." An example might be the signal constituted by understated elegance (which we often call "exquisite"). It costs to eschew a signal that the signaler *could* afford and that would yield a higher return than the signal that conforms to the signaler's high-value discretion and restraint. Choosing such a signaler signals one's own exquisite taste.

One place I hoped to find a vivid example of such self-referentiality was in cichlids. Male cichlids build useless sand nests in places where the current will wash them away. Interestingly, the Zahavis say, *females* lay eggs in those nests that then get washed away (1997: 31). I wanted to propose this as an example of costly signaling through the selection of a costly signaler; if it were true it would be a good one. Unfortunately the example is, in fact, inaccurate, although presumably the Zahavis could bring others to mind: I checked their reference (McKaye 1991), and after I couldn't find exactly what they said, I e-mailed McKaye, who says that they misread him (email to me August 4, 2005). However Bliege Bird, Smith, and Bird (2001: 13) do note that among the Meriam, spear-fishers have higher social status, but provide less food to their families than collectors, so that any person who selects a spear fisher as a mate is also paying a cost. Boone (1998) makes an argument for human costly signaling of altruism that might also provide a reason (though he doesn't say so) to bear the costs of choosing an apparently less-viable partner. Some types might do less well in short-term interactions but survive calamitous population crashes far better through cooperation than might more immediately "rational" types. Boone thinks that human altruism might be a costly signal, paid by a depression in immediate viability, of the capacity to survive a crash through the ability to marshal cooperative resources (see also Fehr and Henrich 2003). I speculate that this would mean that the costs of such signals are designed to be particularly attractive to those who would bear the costs of selecting such signalers because they too are better suited to survive a crash.

The Zahavis are themselves skeptical of group selection (1997: 13–14, 131–132). But their account of the benefits true altruism yields in prestige among the babblers they have been studying for thirty-five years suggests that their way of thinking is consistent with the group selection models to be found in Sober and Wilson (1998), and in Fehr (2002), Boyd and Richerson (2005), Henrich et al. (2006) and Geoffrey Miller (2000). Indeed, in an e-mail to me Amotz Zahavi agreed that (in his words) "females pay a cost when they prefer a male that risks his life for advertisement. They may lose a collaborator to help them rear offspring and their sons are also likely to risk their life advertising their quality" (August 11, 2005). In doing so they are certainly costly signalers in an *inclusive* sense, since they are entrusting their genes in the next generation to sons who are costly signalers and to daughters who prefer costly signalers just as they have. But Zahavi is quick to insist on how "natural selection is calibrating the investment of both males and females to be such that on average advertisers are gaining more from advertising than they spend on it." Nevertheless, he accepts my speculative formulation that females too may be advertising the costs they can afford by picking males who can afford to advertise, and I think some of the evidence the Zahavis give of males being difficult in order to make sure the females are really interested is also consistent with this. Searcy and Nowicki (2005) are particularly attentive to what they call "receiver-dependent costs" (e.g., pp. 14–15, 156, 186), that is, costs imposed by the fact that the signal is received, both by its intended recipient and by potential interceptors. It's easy to see how these costs, which increase the value and honesty of the signal, could therefore allow a signaler to use the recipient as a signal *to* the recipient. Sometimes these costs will include unintended or secondary receivers, whose possible existence is therefore part of the cost, as when nurslings risk attracting the attention of predators in begging from their parents (Searcy and Nowicki 2005: 203). It may even be that the recipient of the signal has the attractive quality. Thus, as I'll argue in Chapter 2, trophy spouses may be signals to themselves: "My spouse can afford to show me off, despite the risks posed by rivals who might court me; the more attractive I am the more proof I have of my spouse's fitness, which is what I seek." In one complex scenario Rowell, Ellner, and Reeve (2006) consider a situation where a male signals in such a way as to attract both females and rivals (p. E189). Though they don't say so, the fact that rivals are likely to eavesdrop on the signal may be part of its cost, since (as they show) it enters into the females' calculations as to whether to respond to the signals (which may be dangerous), that they may also have the opportunity to mate with the rivals. (I hasten to add that while most examples of sexual selection tend to turn on female choice, this is partly an artifact of the obvious macroscopic examples, mainly birds, and that in other organisms, including some butterflies, the choice goes the other way.)

96. Frans de Waal (2005), among others, accepts the idea that altruism makes sense as a default mode in Paleolithic times, on the assumption that most people one encounters will either be close kin or at least likely to be in a position to reciprocate when necessary. But the prevalence of one-shot interactions, along with the selective pressures that would weed them out if Trivers (1971) were right, makes it likely that altruism, even in one-shot interactions, was not a default but an actual benefit. The idea of altruism as advertising fitness helps to explain this.

97. In a dense but beautifully lucid exposition of the ultimatum game, Nowak, Page, and Sigmund (2000) show how a reputation (which is, in fact, a signal, as Robert Frank [1988] points out) for accepting low offers in the ultimatum game will make that rational strategy a losing one in repeated interactions within a population whose members have a positive chance of knowing that reputation. Also, they show how in such a population making and accepting only high offers turns out to be the optimal strategy, the one that will thereby be selected for, whether by genes or by culture, and the one that will therefore lead to the most reproductive success.
98. Camerer (2003: 66–68) reviews some evidence tending to show that altruistic behavior in ultimatum games develops as children grow older, which also suggests a cultural cofactor in the evolutionary biology of altruism (cultures have to turn it on) as against the idea that it's a maladaptation. So too does the evidence he summarizes (pp. 68–74) on the factors that predict differences between cultures in ultimatum games: essentially, the more cooperative a culture is and the more it's achieved "market integration," the less "rational" the moves in an ultimatum game. Camerer notes the fascinating fact that there are some cultures that make hyperfair offers—offers of more than 50 percent of the stakes, and that sometimes those hyperfair offers are rejected, according to the logic of the potlatch by which giving signals status and accepting signals subordinance. (This would be a version of the costs born by the receiver of a signal, which I mention in note 96.)
99. It should be noted, however, that R. A. Fisher (1915) regards ethical behavior and the evaluation of the ethical motives of others—including altruism—as the thing we become most concerned with in selecting our mates. Fisher (like Trivers) has a theory of reciprocal altruism in germ, and also a theory that sexual selection can get you beyond reciprocal altruism and select for genuine altruism.
100. For a version of such behavior among nonhuman primates, see Flack, de Waal, and Krakauer (2005); this article is particularly interesting because its authors are skeptical of the idea of costly signaling but concede its relevance here.
101. Turner (1982) talks about art and rituals of redress; nothing that I say is inconsistent with his interests, but I am interested in narrative's relation to every day human interaction, not in what sacralizes it.
102. The fact that the characters are most likely doubled adds to the centrality of vicarious experience here. Each is separated from him- or herself in imagining the separated, external erotic *object* (not subject) that would satisfy those he or she loves. I take this as a special case of the experience of

the disguised Odysseus has when he hears his own story recounted by the bard Demodokos, and of Scrooge when he sees his young self with tenderness and compassion. I'll mention another example momentarily in considering Thomas Nashe's account (1958) of the Talbot scenes in *1 Henry VI*.

103. This is Darwin's formulation (about Odysseus's reunion with Telemachus, and also about Penelope's with Odysseus: "in such cases we may be said to sympathize with ourselves" [2004: 215]), and Darwin regards sympathy as an irreducible and different emotion from the emotion sympathized with. A similar idea is already in Plato (1997), who thinks that sympathy with fictional characters is baneful because it will *lead to* self-pity. Sympathy comes first. Plato's general argument is no doubt partly directed against what Aristophanes has Aeschylus say in *The Frogs* about how an audience is induced to measure itself against and rival a sympathetic hero (1943: ll. 1042–1043).
104. Philip Fisher (2002) calls this "volunteered affect." We will return to this issue in Chapter 4.
105. The difference between how we feel for the sake of others and how we feel about our own positions is one that Smith (2004) and Hume (1978) describe well. Noël Carroll (1998) seems to have this difference in mind and then neglects it in his argument against identification, an argument whose telling points I agree with. But like Tooby and Cosmides (2001), he seems too quick to assimilate emotions to a kind of propensity for action such as we *would* have if we were the person or character that we see, and this I disagree with, as I'll show in Chapter 4.
106. "My son / Came then into my mind, and yet my mind / Was then scarce friends with him. I have heard more since" (Shakespeare, *King Lear* 2001: IV.i.35–37).
107. On gossip as a mode of punishment and enforcement of norms, see Axelrod, *The Complexity of Cooperation* (1997). This clearly fits with the kind of gossip to be found in Zola, especially *L'assomoir,* and also the disciplinary features of gossip analyzed by such Foucaultian critics as D. A. Miller in *The Novel and the Police* (1988). Brian Boyd (2001) assumes gossip is essentially merely an exchange of information.
108. Brian Boyd (2001) talks about storytellers as getting a useful reward in their audience's attention. He and Sugiyama (2001) also take it that storytelling imparts information and belongs to a Trivers-like structure of direct reciprocal altruism. Tell others important things, and they'll return the favor. Perhaps, but I wish to see it as belonging also to the strong reciprocity or true altruism I have been expounding. Most people would rather tell than listen to a story (or a joke), which suggests that the psychological incentive overcomes the costs of telling, and to get someone to prefer to lis-

ten the teller has to be very impressive indeed. Javier Marías's extraordinary sequence of novels, *Your Face Tomorrow* (2005), contains extended meditations on the cost to their narrator both of narrating and of listening.

109. In Barash (2005: 2) the authors make more or less the same mistake as early Freudian literary criticism—they treat literary characters as motivated by the same things that motivate real humans, rather than as representations to whom real humans react. It's our reactions that psychology can analyze, not the actions of literary characters. "*Othello,*" they say, "speaks to the Othello within everyone: our shared human nature." "Speaks to" covers a multitude of problems. Why and how does it speak to that? What does it say? "Speaks to" is another way of saying "makes us identify with," and is no more helpful. If, as I believe, we do not identify with literary characters, their verisimilitude needn't and usually doesn't involve the same natural psychological constitution as that of their audience.
110. It was, until recently, the law of every state in the United States that the husband of any woman who gave birth to a child was legally that child's father. Cheerleading science journalists have said that such a law is anachronistic now, because of DNA testing. I think it's a great law and should not be repealed.
111. Indeed, all the behavioral experiments seem to bear this idea out. Very few people always defect; very few are strong reciprocators in every circumstance. We are all a mixture of self-interest and genuine altruism, and different situations, dramatis personae, times of life, will appeal to different aspects of our character. For more on individual variability of response, as well as its limits, see Camerer (2003: 45–46). As ever, Hume anticipates this argument when he writes: "It is sufficient for our present purposes, if it be allowed, what surely without the greatest absurdity cannot be disputed, that there is some benevolence, however small, infused into our bosom; some spark of friendship for human kind; some particle of the dove kneaded into our frame, along with the elements of the wolf and serpent. Let these generous sentiments be supposed ever so weak . . . they must still direct the determinations of our mind, and where everything else is equal, produce a cool preference of what is useful and serviceable to mankind, above what is pernicious and dangerous" (1975: 271).
112. For more on the stability of cooperative types see Kurzban and Houser (2005).
113. Sidney cites a similar paradox in tragedy: "Plutarch yieldeth a notable testimony of the abominable tyrant Alexander Pheraeus, from whose eyes a tragedy, well made and represented, drew abundance of tears, who without all pity had murdered infinite numbers, and some of his own blood: so as he, that was not ashamed to make matters for tragedies, yet could not

resist the sweet violence of a tragedy" (1971: 45). This refers to Alexander of Pherae's anger at his own weeping, which caused him to leave the theater lest he "be seen to weep over the sufferings of Hecuba and Polyxena" (Plutarch, Moralia: "On the Fortune of Alexander") when he was himself killing others, so that he is aware both of his own altruistic responses and that he is being monitored by others even as he watches the tragedy. Plutarch says that he came close to killing the actor who drew tears from him, not so perverse a reaction when seen as an aspect of altruistic punishment. Proust remarks (also about actors and their jealous rivalries) that except in cases of actual sadism, "the nasty man thinks he's punishing someone nasty" (1987: 2.472).

2. Signaling

1. Another cost might be the abandonment of the element of surprise, to which we'll return momentarily, as when leopards make themselves visible well before they are within pouncing range.
2. Most bluffing has little to do with winning the actual hand being played; most good players bluff to try to establish a false reputation as a bluffer among worse players (see Van Neuman and Morgenstern [2004]), counter to the generally costly and honest signaling most betting really is. Bluffing is a technique that works mainly against poor players. (A good player has to be willing to call a bluff against bad players, that is, to probe the honesty of the signal. Signals that are too dishonest become prohibitively or at least destructively costly, but even, or especially, against bad players good players will almost never bluff except strategically.) The fitter the players a tournament has selected for, the more honest their signals to one another will be. Poker presents a formidably but attractively complex case of signaling among people, which I won't follow up, except to note that the distribution of *hands* (provided by "nature") is a question of chance and conforms after a series of games to the law of averages. Betting therefore signals not only the quality of the hand that a player has in any particular round, but also the skill of the player. A good player can cheat a little in this signaling by deceptively seeming more deceptive than he or she is, that is, by bluffing in order to establish a reputation as a bluffer. But this is itself a costly strategy (bluff too often and you lose a lot of money), so that the limits within which a good player can signal deceptively are very narrow, and most of his or her signaling will be honest. It's in that very narrow range of deception within a generally honest signal that good players compete, just as in ambiguous and nearly matched cases animals actually fight.
3. For this general argument, see Robert Frank (1988).

4. See the Zahavis' (1997: 10–12) description of honest signaling in boxing in their account of "signaling to prey," for example: "Early in a bout, many of a boxer's moves are made to determine the opponent's agility in defense. We believe these exercises help the fighter—or predator—not by misleading but rather by offering precise information about the attacker's repertoire. The more reliable the information provided by the attacker, the more he finds out."
5. Some guppies court by using spot colors highly visible to the guppies they're courting but inconspicuous to predators, because the predators see the guppies from different angles and are sensitive to different wavelengths of light (Guilford and Dawkins 1991: 3). Winking is a typical human signal designed to be invisible to at least some observers: you wink with the eye that can be seen by the person you intend to see it, and not by the person from whom you intend to keep it hidden.
6. As I suggested above, this may be one reason it would be very hard to design an experiment to measure accurately our response to altruistic punishers, since a slight difference in how we see them makes an enormous difference for whether we approve of them.
7. See Bliege Bird, Smith, and Bird (2001: 9) on storytelling as mechanism for conveying the costly signaling of spearfishing (a highly inefficient mode of food production) among the Mer.
8. Inadvertent honesty requires great delicacy on a narrator's part if we are not to begin to feel positively toward its signaler. This is a correlate of the self-limiting extent to which we can take pleasure in the discomfiture of someone feeling agonizing remorse.
9. Remember the definition of "strong reciprocation" from Chapter 1: disinterested and truly altruistic response to observed behavior in an interaction independent of the observer.
10. I follow the Folio reading here; the Quarto has *they'll,* but I think the Folio's revision is intentional and shows how it is that Edgar comes into Gloucester's mind at this moment, as we have seen (Chapter 1, note 106).
11. See the Zahavis (1997: 112) on "Testing by Imposition": "The only way to obtain reliable information about another's commitment is to impose on that other—to behave in ways that are detrimental to him or her. We are all willing to accept another's behavior if we benefit from it, but only one truly interested in the partnership is willing to accept an imposition . . . All mechanisms used to test the social bond involve imposing on partners."
12. This is not voluntaristic "wishing." Most people can detect willful weeping—what we call sulking (which is itself a costly signal). This may be why Hamlet insists that his more or less unconvincing tears ("the fruit-

ful river of the eye") cannot denote him truly. But they probably do—they denote the willfulness of his weeping when we first meet him.

13. Tears, Darwin (2004) says, are a byproduct of crying; they express emotion but did not evolve to do so. The fact that they blind the crier may have evolved or become adapted to use as a costly signal. See Hauser (1996: 469) and Hauser (2006: 247–248).
14. I should stress that I am describing how Shakespeare represents Antony on stage and in his lines, a representation untouched by the questionable things we know Antony does do, like the offstage and nearly unremarked murder of Lepidus. The costs Antony bears are costs to his rational self-interest; Shakespeare makes it clear that his self-interest is titanic and often massively self-serving, which makes Antony's onstage willingness to pay the costs all the more affecting. As in any tragedy, the onstage figure is the only one who counts for us as a person.
15. For a grim version of this logic see Cynthia Ozick's story "The Shawl" (1990).
16. I am not offering a psychological or critical account of Hal here; the qualities he communicates are public and accepted by the public (in the play and through it), whatever his own more critically interesting motives are. I address these motives in Flesch (1992).
17. In fact, the benefits don't always outweigh the costs, though they may on average. There's some very recent evidence that the costs are sufficient to lead to a kind of feast-or-famine outcome for offspring, poor outcomes being more likely especially if genes that signal fitness in one sex are passed on to offspring of the other sex. This tendency, if it's widespread, may be good for the desideratum of maximizing variation within a group. See "Most Desirable Mates May Not Sire Most Prolific Offspring" (*Science News* 2006) and Pischedda and Chippindale (2006).
18. For some citations supporting this analysis, see Chapter 1, note 95.
19. It may be helpful—or not—to think of this argument as a generalization and extension of the dialectic of lordship and bondage in Hegel's *Phenomenology of Spirit* (1977), where master and slave each find themselves only through the other who represents them. Here the publicity of what each member of the communicating couple is to the other gets at an irreducible element of vicarious experience—experience of the other as public and publicized and as experienced by the public—that in Hegel is confined to the dyadic relationship.
20. Again: as the female cichlid would advertise her own quality by laying eggs in the costly folly the male has built for her, where both know the eggs will be swept away. While many may admire, only those willing to pay these costs would mate with each other.

21. George Eliot seems to be ringing some ironic changes on this scenario in *Daniel Deronda* (1999) when she has Gwendolyn lose her money gambling. What we hope will be attractive to Daniel turns out not to be; he'll pay the price for loving the suicidal Mira but not the self-destructive Gwendolyn, who, presented at first as a different kind of heroine, seems to be signaling in just the costly way that would make her into the heroine.
22. This kind of spectatorship is based on the way we scrutinize one another to ferret out second-order free-riders. Here I am interested in isolating the spectatorship itself, as narrative does. Such spectatorship need not be in every situation the second-order free-riding that it sometimes is. We're not *always* expected to be altruistic punishers; expectation will be pegged to the costs of punishment. Sometimes we're required not to be, since altruistic punishment can be a sign of high status. The higher the costs, the fewer the organisms that can bear them. If another (possibly higher status) altruistic punisher is doing the work we need only monitor and approve the costly signal that that work constitutes. We *are* always expected to be impressed by altruistic punishment and to approve of it when it's fair, that is, successful and felicitous.
23. See also Ainslie (2001: 3–70, especially pp. 24–26).
24. I stress again that this is a very quick simplification of the subtle and groundbreaking argument in Robert Frank's seminal book (1988).
25. For an argument nicely consistent with mine, although put to polemical ends, see O'Farrell (1997: 49–50).
26. See Cavell's great essay "The Avoidance of Love" in Cavell (2003).
27. See Nagel (1974).
28. On emotions as solving commitment problems, see Schelling (1978) and Robert Frank (1988).
29. Philip Fisher (2002) has different reasons for describing volunteered emotion. Like Dr. Johnson, he tends to understand them as the emotions appropriate to a situation that we know more fully than those we pity.
30. Much of the conclusion to the second *Enquiry* (1975) anticipates in its account of trust, reputation, and the risks of pure selfishness, Frank's avowedly Humean *Passion within Reason* (1988).
31. Nagel is intensely interested in the experience of subjectivity. One powerful move that he makes is to divorce that experience from the perspectivalism is seems to imply. He argues that "something essential about me has nothing to do with my perspective and position in the world . . . I throw TN [i.e., the empirical Thomas Nagel] into the world as a thing that interacts with the rest of it, and ask what the world must be like from no point of view in order to appear to him as it does from his point of view" (1974: 62). I am highly sympathetic to this argument, but sympathetic as

well to the Wittgensteinean reminder (about which Nagel [1974: 37] is somewhat ambivalent) that we live publicly in the public world, that *most of the time* we don't ask these questions at all, absorbed unselfconsciously into the general world of which we are a part. A phenomenological, psychoanalytic, and narrative account of vicarious experience, which I hope to attempt in a later book would have to go into some of the evidence from experimental psychology that perspectival representations, like paintings, photographs, and films, rely far less on the so-called "point assigned by perspective" (Damisch 1994) or perfect viewpoint than geometrical theorists have imagined (see Kubovy [1988] for a fascinating account of the experimental evidence). There's a large spatial region, not a single ideal point, from which we can see things perspectively, because we see not from a particular point but from a large volume of space into which we are absorbed. The sense data come to us as concentrated into a pointlike location (the retina) but we *see* not from our own private locations behind our eyes, but from the open available *general* space around us. See also Koch's (2004) chapter on seeing and night-walking.

32. See also Plato's *Republic* 606c on "the laughable" (1997).
33. Even Dr. Johnson, who thinks that the emotions tragedy come from imagining our own selves in the tragic situation we see depicted, describes a vicarious relation to our own feeling:

> If there be any fallacy, it is not that we fancy the players, but that we fancy ourselves unhappy for a moment; but we rather lament the possibility than suppose the presence of misery, as a mother weeps over her babe, when she remembers that death may take it from her. (*Preface to Shakespeare* 1765)

But the mother is weeping over her baby as much as over herself, or if she is luxuriating in her own possible grief, she is doing so as she would not if the grief were actual and she were in fact mourning her dead child. The medium of possibility is exactly that of the nonactual. Weeping over how she might feel is very different from weeping out of actual feeling, and produces an affect proper to vicarious experience, even of one's own possible experience.

34. As Feste argues against the mourning Olivia in *Twelfth Night*.

3. Storytellers and Their Relation to Stories

1. It is a costly signal, since his communication of the contrivances he doesn't put into practice risks her hatred, as the sequel to this passage shows: "Beg of her not to hate thee for the communication; and assure her

that thou gavest them up from remorse, and in justice to her extraordinary merit; and let her have the opportunity of congratulating herself for subduing a heart so capable of what thou callest *glorious mischief*. This will give *her* room for triumph; and even *thee* no less: she for *hers* over thee; thou for *thine* over *thyself*" (Richardson 1932: 2.319). Belford urges the idea that the changes that Lovelace might allow to his contrivances would be a mode of altruism in which she would be the costly signal constituted by his own ability to triumph over himself, even as that very fact serves to woo her for the altruism by which she triumphs over his own propensity for mischief.

2. Even other authoritative narrators must defer to the original contriver, as when Stephen King and John Irving begged Rowling not to kill Harry in the not-yet-released last volume of the *Harry Potter* series.
3. Fan fiction, practiced since the 1970s, now has thousands of practitioners, thanks to the Internet. Fan fiction doesn't actually think of itself as changing the world of the characters it enlists; it's fiction in a conceptually subjunctive mood, and so allows for the expression of audience wishes, rather than conceptually indicative "fictional facts." I mention it here because the rules and terms of judgment that its participants have established are interesting in themselves and give a good sense of the canons of plausibility and of other modes of generic or general understanding that all audiences bring to the fictions they attend to. A lexicon of the generic terms of fan fiction constitutes an ad hoc lexicon of narrative. The way people judge one another's work is continuous with the way they judge the work of the authoritative fiction they embroider. Such terms as H/C (hurt/comfort, a way of getting characters together), UST (unresolved sexual tension), OTP (one true pair, the central erotic relationship), and especially OOC (out of character), suggest the fans' sense of possibilities and limits within the fiction they monitor; the possibilities and limits in which they work also constrain the authors of the authoritative fiction, even if such constraints are more tacit in our interactions with the original authors or works. The *way* that in fan fiction everyone's a critic suggests the ways in which we are involved in an approving or disapproving fashion with what storytellers do to their characters. (The terminology of fan fiction calls its producers "active fans.") A good and entertaining lexicon may be found at *The Fanfiction Glossary*.
4. This line should be contrasted with Walton's (1990) claim that we imagine *ourselves* seeing what we visualize.

   You might argue that it's trivial to say that no one sees what's verbally described. But the narrator (not Watson by the way, but an omniscient one here) nevertheless calls it a *sight*, so that he has some access to it as a sight and not merely as description. What's being described (in the indicative) is a sight, something seen therefore, though no one is there to see it. (The buzzards see it, but not its strangeness, so the strange sight doesn't refer to their seeing.)

5. William Goldman said movies should give the audience what it wants but not in the way it expects it.
6. Consider also the way that Lily Bart has to worry about the various judgments made by those observing her and observing others observing her and one another in *The House of Mirth* (1994). But she is not so artful as Maggie. Zunshine (2006) discusses the extent to which it's possible to keep track of what people know about what other people know about what other people know about.
7. This scene also presents a beautiful example of differential signaling for a multiple audience. What the Op is doing is both focused and broadcast. Tom-Tom reads and responds to the information differently from Foley and Linehan, and Jack differently still. Hammett has conceived a scene that illustrates perfectly the signaling situations envisioned by Rowell, Ellner, and Reeve (2006), in which they show how "a signaler can benefit from using a mixture of honesty and dishonesty when faced with dispositionally distinct receivers." In describing a mating scenario they show how such dispositional distinctions allow for a single signal's being a "partial deception," based on the responses of its receivers: "If the signal is 'honest enough,' then the females should act as if it were true, so the signaler does not need to be totally honest. If the signal is 'deceptive enough,' then satellites should stay out, so total dishonesty is not necessary." The situation they describe is one in which the signaler signals his desire to mate, but then sometimes ambushes and attacks all responders. "The key factor is that satellite males and females exhibit differing levels of tolerance toward the cost of being injured. Female willingness to settle for matings with satellite males improves the likelihood that a partial ambush strategy will not cause females to avoid the territory" (Rowell, Ellner, and Reeve 2006: E189). "Receiver psychology" (as Guilford and Dawkins call it [1991]) makes all the difference (as with the red cape and the bull).
8. This would be to make John Grady the haruspex of his own sacrifice, reading the future in his own viscera. McCarthy returns to this notion of grim and hopeless prophecy in *The Road* (2006), set in a final nuclear winter where, as a character says, there is not one prophecy of doom that has not come true.
9. Remember the Zahavis (1997: 10–12) on boxing.
10. Here it's worth quoting J. M. Coetzee's (2004) essay on Philip Roth's novel *The Plot against America* (2004). What in real life—the life that the real Philip

Roth lives—makes plausible and therefore truthful the language of autobiography that presents the semifictional Roth as scarred and ruined by the American period of fascism delineated in that novel's fictionalized history? "The only answer that seems to make sense . . . is that the scar is Jewishness itself, but Jewishness of a particular etiology: Jewishness as an outsider's idea—and a hostile outsider's at that—of what it is to be a Jew, forced upon the growing child too early, and by means that, while they might not be extreme in themselves, might easily . . . become extreme" (Coetzee 2004: 5). Although Coetzee doesn't quite say so, that hostile outsider's view is what Roth the writer also faces as a result of the novels that themselves respond to such a view, and therefore his writing about damage in *The Plot against America* responds to real damage, to himself as a person, and as a writer.

11. Paul de Man's great essay on Kleist (1984) compares Leiris's essays on writing compared to bullfighting to Kleist's on the marionette theater.
12. "An indescribable delight always springs forth from the great books, even when they present things that are ugly, desperate, or terrifying" (Deleuze 1977: 147).
13. I am, of course, aware of the relevance of this description to Harold Bloom's theory of poetic influence (1973). Influence is altruism ambivalently resented by the beneficiary.
14. Beckett (1956) describes such a moment in *Malone Dies,* where one character waits while another read the letter she has written him: "While he read Moll held a little aloof, with downcast eyes, saying to herself, Now he's at the part where, and a little later, Now he's at the part where, and so remained until the rustle of the sheet going back into the envelope announced that he had finished" (p. 260).
15. In his *Autobiography* (1882) Trollope describes the motives that induced him to kill off "my old friend Mrs Proudie." As he is writing at his club he overhears two clergymen who do not know him, both of whom happen to be reading parts of "some novel" of his in magazines, complain that he introduces the same characters too often.

> Then one of them fell foul of Mrs Proudie. It was impossible for me not to hear their words, and almost impossible to hear them and be quiet. I got up, and standing between them, I acknowledged myself to be the culprit. 'As to Mrs Proudie,' I said, 'I will go home and kill her before the week is over.' And so I did. The two gentlemen were utterly confounded, and one of them begged me to forget his frivolous observations.
>
> I have sometimes regretted the deed, so great was my delight in writing about Mrs Proudie, so thorough was my knowledge of all the little shades of her character. (pp. 108–109)

Trollope pays the high cost of killing off a character he loves because it is the proper thing to do. In doing so he also shows up the clergymen, and so succeeds in a bit of altruistic punishment, in a highly gentlemanly fashion—one moreover which yields the later narrative of his *Autobiography.*

4. Vindication and Vindictiveness

1. Browning's *The Ring and the Book* (1991) is perhaps the most brilliantly sustained staging in English of a competition between what all sides regard as their vindicable behavior and the vindictiveness shown by the other side. In particular the Pope and Guido each think the other purely vindictive. What allows us to orient ourselves is Browning's own guarantee of Pompilia's innocence, from which every other judgment as to the distinction between vindication and vindictiveness flows. It is the fact that Guido himself finally acknowledges her innocence, and seeks to punish her for *that,* that confirms us in seeing his altruistic punishment as hateful vindictiveness and spite:

   > There was no touch in her of hate:
   > And it would prove her hell, if I reached mine!
   > To know I suffered, would still sadden her,
   > Do what the angels might to make amends!
   > Therefore there's either no such place as hell
   > Or thence shall I be thrust forth, for her sake,
   > And thereby undergo three hells, not one—
   > I who, with outlet for escape to heaven,
   > Would tarry if such flight allowed my foe
   > To raise his head, relieved of that firm foot
   > Had pinned him to the fiery pavement else!
   > So am I made, "who did not make himself:"
   > (How dared she rob my own lip of the word?)
   > Beware me in what other world may be!
   > Pompilia, who have brought me to this pass! (XI: 2088–2101)

   I cite Browning here because Guido gives a powerful instantiation of altruistic punishment, or spite, as an endowment of nature ("So am I made").
2. Thackeray's narrator, who tells us that he reads notes over Becky's shoulder and also hears the events he recounts from people who knew the principles, like Hawthorne's in *The House of the Seven Gables:*—

   > It still lacked half an hour of sunrise, when Miss Hepzibah Pyncheon—we will not say awoke, it being doubtful whether the poor lady had so much as closed her eyes during the brief night of midsummer—but, at

> all events, arose from her solitary pillow, and began what it would be mockery to term the adornment of her person. Far from us be the indecorum of assisting, even in imagination, at a maiden lady's toilet! Our story must therefore await Miss Hepzibah at the threshold of her chamber; only presuming, meanwhile, to note some of the heavy sighs that labored from her bosom, with little restraint as to their lugubrious depth and volume of sound, inasmuch as they could be audible to nobody save a disembodied listener like ourself. (Hawthorne 1913: 46)—

and James's in *The Golden Bowl:*—

> Adam Verver, at Fawns, that autumn Sunday, might have been observed to open the door of the billiard-room with a certain freedom—might have been observed, that is, had there been a spectator in the field. The justification of the push he had applied, however, and of the push, equally sharp, that, to shut himself in, he again applied—the ground of this energy was precisely that he might here, however briefly, find himself alone. . . . The vast, square, clean apartment was empty, and its large clear windows looked out into spaces of terrace and garden, of park and woodland and shining artificial lake, of richly-condensed horizon, all dark blue upland and church-towered village and strong cloudshadow, which were, together, a thing to create the sense, with everyone else at church, of one's having the world to one's self. We share this world, none the less, for the hour, with Mr. Verver; the very fact of his striking, as he would have said, for solitude, the fact of his quiet flight, almost on tiptoe, through tortuous corridors, investing him with an interest that makes our attention—tender indeed almost to compassion—qualify his achieved isolation. (James 1987: 129)—

likes to make jokes about that status of his presence and knowledge of events in the world whose story he tells, but the jokes point the moral that he doesn't live in that world. James's splendid formulation of the tactful and tender way our attention—that of a disembodied spectator like ourself, as Hawthorne might put it—qualifies but doesn't destroy Adam Verver's isolation is a little parable about our self-forgetfulness in reading.

3. Borges's story on Jesus and Judas tells how the true savior who abased himself to save us all through establishing the drama of the Passion was not Jesus but Judas. Jesus was his means, but the punishment that Judas accepted, and by which he redeemed us, was to be despised and unforgiven by everyone throughout history (Borges 1969), even, we could say, by the real Borges himself, whose light, ironic parable doesn't seriously propose the scenario it entertains.

4. Modern versions of such witnesses of vindication might include such posthumous narrators as Joe Gillis (in Brackett, *Sunset Blvd.*, 2002) and Lester Burnham (in Mendes, *American Beauty*, 2000).
5. The phrase "provoking merit" is somewhat hard to parse, but most commentators who notice it treat "provoking" as a kind of middle-voice participle, a merit called forth by provocative behavior. This may be, but perhaps we should understand it this way: The kind of merit shown by Edgar in response to the reprovable badness Cornwall accuses Gloucester of might naturally turn people against Edgar, just as Cordelia's merit might naturally attract animosity. In general, characters in the play are provoked by what they might also acknowledge upon reflection is merit; the merit is that of the altruistic punisher who pays to punish—pays by provoking dislike and risking mistrust in others. Think of Captain Queeg in *The Caine Mutiny* (Wouk 1951), and the passionate defense of Queeg's altruism that Wouk finally puts in the mouths of those very lawyers who (altruistically enough) punish *him*. At any rate, I like the phrase as capturing the sense that altruistic punishment can be *vexing* to those who observe it; the way it vexes others is one of its costs.
6. P. F. Strawson's great essay on "Freedom and Resentment" (2003) distinguishes between the "reactive attitudes," like resentment, which assume that the other is a free agent, and therefore also someone we wish to make understand something, and the "objective" attitude in which we treat the other in a fashion calculated to maximize our own selfish goals and not as a full and normal human being. Strawson denies that there is any version of determinism that could make most people programmatically and generally deny the humanity of others, and thus make us abandon the interpersonal desire to make ourselves understood to others. His "resentment" captures an attitude that I would like to bring out in both vindictiveness and in the desire for vindication. He distinguishes between treating others purely objectively, as obstacles or instruments to our own goals, and treating them, as I would call it, vicariously: "If your attitude towards someone is wholly objective, then though you may fight him, you cannot quarrel with him, and though you may talk to him, even negotiate with him, you cannot reason with him" (p. 79). The equivalent of the resentment one feels when one is the object of what one reacts against is *indignation* when one reacts on behalf of someone else, and Strawson says that resentment may be thought of as involving expectation of indignation on the part of observers if they could see what the person resented is up to, just as indignation is an expectation of resentment on the part of the victim of the behavior one is indignant at witnessing.
7. The Quarto has "Let him *first* answer that"; by removing it (if he did)

Shakespeare makes the strategic importance of the question even more prominent.

8. The *sir* is a Folio addition/revision. Here is not the place to go into the vexed and important issue of Shakespeare's revisions of *King Lear* (the fact of which I generally accept), except to say that I regard their value as usually showing *not* his rethinking but his reexpressing and intensifying the movement and power of the play. Revision very rarely changes a work, which has a holistic quality, as Shakespeare knew (and as I argue elsewhere), so one's first assumption about revision should be that Shakespeare decided on what he thought was a better way of saying what he'd said before.
9. For a psychoanalytic account of fantasies of vindication in *King Lear,* see Rosenberg (1972: 333–334).
10. As always, Cormac McCarthy tests some of the limits of narrative in *No Country for Old Men* (2005), where the characters we want most to live are murdered; the murderer makes the last one believe that she has been betrayed by the husband who has sought to save her. He gets away with everything. But he is not surprised to find himself in hell.

Coda

1. I think the end of *The Eumenides* (Aeschylus 1960), with the alliance between humans and punishing Furies, and moreover on behalf of cooperation, has much the same feel.

Index

Harvard University Press is a member of Green Press Initiative (greenpressinitiative.org), a nonprofit organization working to help publishers and printers increase their use of recycled paper and decrease their use of fiber derived from endangered forests. This book was printed on 100% recycled paper containing 50% post-consumer waste and processed chlorine free.